New York State Coach, Empire Edition, Mathematics, Grade 4

Coach™

Triumph Learning®

New York State Coach, Empire Edition, Mathematics, Grade 4
312NY
ISBN-10: 1-60824-073-8
ISBN-13: 978-1-60824-073-9

Contributing Writer: Cindy Frey
Cover Image: The Empire State Building. © Jupiterimages/Comstock

Triumph Learning® 136 Madison Avenue, 7th Floor, New York, NY 10016

Printed in the United States of America.

10 9 8

Table of Contents

			New York State Math Indicators	Common Core State Standard(s)
Letter to the Student 6				
Test-Taking Checklist 7				
New York State Math Indicators Correlation Chart 8				
Strand 1	**Number Sense and Operations** 13			
Lesson 1	Place Value 14		4.N.1, 4.N.2, 4.N.4, 4.N.5	4.NBT.1
Lesson 2	Compare and Order Whole Numbers 20		4.N.3, 4.A.2	4.NBT.2
Lesson 3	Round Whole Numbers 26		4.N.26	4.NBT.2, 4.NBT.3
Lesson 4	Add and Subtract Whole Numbers 31		4.N.5, 4.N.14	4.NBT.4
Lesson 5	Multiplication Facts 37		4.N.16	4.OA.1, 4.OA.2
Lesson 6	Properties of Multiplication 43		4.N.6, 4.N.13	4.NBT.5, 3.OA.9 ●
Lesson 7	Multiply by One-Digit Whole Numbers 48		4.N.18	4.NBT.5
Lesson 8	Multiply by Two-Digit Whole Numbers 52		4.N.19	4.NBT.5
Lesson 9	Division Facts 56		4.N.16, 4.N.17	4.OA.1, 4.OA.2
Lesson 10	Divide by One-Digit Divisors 62		4.N.21, 4.N.22	4.NBT.6
Lesson 11	Multiply and Divide by Multiples of 10 and 100 70		4.N.20	4.NBT.5, 4.NBT.6
Lesson 12	Fractions 74		4.N.7	3.NF.2.a ●, 3.NF.2.b ●
Lesson 13	Equivalent Fractions 77		3.N.14*, 4.N.8	4.NF.1
Lesson 14	Compare and Order Fractions 84		3.N.15*, 4.N.9, 4.A.2	3.NF.3.d ●, 4.NBT.2
Lesson 15	Add and Subtract Fractions 91		4.N.23	4.NF.3.a, 4.NF.3.b, 4.NF.3.d
Lesson 16	Decimals 96		4.N.10, 4.N.11	4.NF.6
Lesson 17	Compare and Order Decimals 103		4.N.12, 4.A.2	4.NBT.2, 4.NF.7
Lesson 18	Add and Subtract Decimals 109		4.N.25	
Lesson 19	Fraction-Decimal Equivalencies 115		4.N.24	4.NF.5, 4.NF.6
Lesson 20	Solve Real-World Problems 121		4.N.15	4.OA.3
Lesson 21	Use Estimation to Check Answers 126		4.N.27	4.OA.3
Strand 1 Review 131				

● Standard covered at another grade level.   * Grade 3 May–June Indicators   ** Grade 4 May–June Indicators

Strand 2   **Algebra** 139

Lesson 22   Open Sentences with Equations 140   4.A.1   4.OA.3
Lesson 23   Open Sentences with Inequalities 143   4.A.1–4.A.3
Lesson 24   Number and Geometric Patterns 148   4.A.4   4.OA.5
Lesson 25   Input-Output Tables .155   4.A.5

Strand 2 Review 161

Strand 3   **Geometry** 167

Lesson 26   Lines 168   4.G.6**   4.G.1
Lesson 27   Angles 175   4.G.7**, 4.G.8**   4.G.1
Lesson 28   Two-Dimensional Figures 181   4.G.1, 4.G.2   4.G.1, 4.G.2
Lesson 29   Perimeter and Area 187   4.G.3, 4.G.4   3.MD.5.a●,
3.MD.5.b●,
3.MD.6●,
3.MD.8●

Lesson 30   Three-Dimensional Figures 193   4.G.5

Strand 3 Review 197

Strand 4   **Measurement** 203

Lesson 31   Customary Units of Length 204   4.M.1–4.M.3   4.MD.1, 4.MD.2
Lesson 32   Metric Units of Length 211   4.M.1, 4.M.2   4.MD.1
Lesson 33   Metric Units of Mass 217   4.M.4, 4.M.5   4.MD.1
Lesson 34   Metric Units of Capacity 222   4.M.6, 4.M.7   4.MD.1
Lesson 35   Make Change 225   4.M.8   4.MD.2
Lesson 36   Elapsed Time 231   4.M.9, 4.M.10

Strand 4 Review 236

● Standard covered at another grade level.   * Grade 3 May–June Indicators   ** Grade 4 May–June Indicators

Strand 5   **Statistics and Probability** 241

Lesson 37   Formulate Questions and Use Surveys 242   3.S.1*, 3.S.2*

Lesson 38   Display Data 247   4.S.3

Lesson 39   Collect and Record Data 253   4.S.1**, 4.S.2**

Lesson 40   Analyze Data and Make Predictions 259   4.S.5, 4.S.6

Lesson 41   Line Graphs 266   4.S.4

Strand 5 Review 271

Glossary 278

Comprehensive Review 1 283

Comprehensive Review 2 301

Punch-Out Tools 319

● Standard covered at another grade level.   * Grade 3 May–June Indicators   ** Grade 4 May–June Indicators

Letter to the Student

Dear Student,

Welcome to *Coach*! This book provides instruction and practice that will help you master all the important skills you need to know, and gives you practice answering the kinds of questions you will see on your state's test.

The *Coach* book is organized into chapters and lessons, and includes two Comprehensive Reviews. Before you begin the first chapter, your teacher may want you to take Comprehensive Review 1, which will help you identify skill areas that need improvement. Once you and your teacher have identified those skills, you can select the corresponding lessons and start with those. Or, you can begin with the first chapter of the book and work through to the end.

Each of the lessons has three parts. The first part walks you through the skill so you know just what it is and what it means. The second part gives you a model, or example, with hints to help your thinking about the skill. And the third part of the lesson gives you practice with the skill to see how well you understand it.

After you have finished all the lessons in the book, you can take Comprehensive Review 2 to see how much you have improved. And even if you did well on Comprehensive Review 1, you'll probably do better on Comprehensive Review 2 because practice makes perfect!

We wish you lots of success this year, and hope *Coach* will be a part of it!

Test-Taking Checklist

Here are some tips to keep in mind when taking a test. Take a deep breath. You'll be fine!

✓ Read the directions carefully. Make sure you understand what they are asking.

✓ Do you understand the question? If not, skip it and come back to it later.

✓ Reword tricky questions. How else can the question be asked?

✓ Try to answer the question before you read the answer choices. Then pick the answer that is the most like yours.

✓ Look for words that are **bolded**, *italicized*, or <u>underlined</u>. They are important.

✓ Always look for the main idea when you read. This will help you answer the questions.

✓ Pay attention to pictures, charts, and graphs. They can give you hints.

✓ If you are allowed, use scrap paper. Take notes and make sketches if you need to.

✓ Always read all the answer choices first. Then go back and pick the best answer for the question.

✓ Be careful marking your answers. Make sure your marks are clear.

✓ Double-check your answer sheet. Did you fill in the right bubbles?

✓ Read over your answers to check for mistakes. But only change your answer if you're sure it's wrong. Your first answer is usually right.

✓ Work at your own pace. Don't go too fast, but don't go too slow either. You don't want to run out of time.

Good Luck!

New York State Math Indicators Correlation Chart

* Grade 3 May–June Indicators   ** Grade 4 May–June Indicators

Indicator	New York State Grade 4 Math Indicators	Coach Lesson(s)
	STRAND 1: NUMBER SENSE AND OPERATIONS	
Number Systems: Students will understand numbers, multiple ways of representing numbers, relationships among numbers, and number systems.		
*3.N.14	Explore equivalent fractions $\left(\frac{1}{2}, \frac{1}{3}, \frac{1}{4}\right)$	13
*3.N.15	Compare and order unit fractions $\left(\frac{1}{2}, \frac{1}{3}, \frac{1}{4}\right)$ and find their approximate locations on a number line	14
4.N.1	Skip count by 1,000's	1
4.N.2	Read and write whole numbers to 10,000	1
4.N.3	Compare and order numbers to 10,000	2
4.N.4	Understand the place value structure of the base ten number system: 10 ones = 1 ten 10 tens = 1 hundred 10 hundreds = 1 thousand 10 thousands = 1 ten thousand	1
4.N.5	Recognize equivalent representations for numbers up to four digits and generate them by decomposing and composing numbers	1, 4
4.N.6	Understand, use, and explain the associative property of multiplication	6
4.N.7	Develop an understanding of fractions as locations on number lines and as divisions of whole numbers	12
4.N.8	Recognize and generate equivalent fractions (halves, fourths, thirds, fifths, sixths, and tenths) using manipulatives, visual models, and illustrations	13
4.N.9	Use concrete materials and visual models to compare and order unit fractions or fractions with the same denominator (with and without the use of a number line)	14
4.N.10	Develop an understanding of decimals as part of a whole	16
4.N.11	Read and write decimals to hundredths, using money as a context	16
4.N.12	Use concrete materials and visual models to compare and order decimals (less than 1) to the hundredths place in the context of money	17
Number Theory		
4.N.13	Develop an understanding of the properties of odd/even numbers as a result of multiplication	6
Operations: Students will understand meanings of operations and procedure, and how they relate to one another.		
4.N.14	Use a variety of strategies to add and subtract numbers up to 10,000	4

* Grade 3 May–June Indicators   ** Grade 4 May–June Indicators

Indicator	New York State Grade 4 Math Indicators	Coach Lesson(s)
4.N.15	Select appropriate computational and operational methods to solve problems	20
4.N.16	Understand various meanings of multiplication and division	5, 9
4.N.17	Use multiplication and division as inverse operations to solve problems	9
4.N.18	Use a variety of strategies to multiply two-digit numbers by one-digit numbers (with and without regrouping)	7
4.N.19	Use a variety of strategies to multiply two-digit numbers by two-digit numbers (with and without regrouping)	8
4.N.20	Develop fluency in multiplying and dividing multiples of 10 and 100 up to 1,000	11
4.N.21	Use a variety of strategies to divide two-digit dividends by one-digit divisors (with and without remainders)	10
4.N.22	Interpret the meaning of remainders	10
4.N.23	Add and subtract proper fractions with common denominators	15
4.N.24	Express decimals as an equivalent form of fractions to tenths and hundredths	19
4.N.25	Add and subtract decimals to tenths and hundredths using a hundreds chart	18
Estimation: Students will compute accurately and make reasonable estimates.		
4.N.26	Round numbers less than 1,000 to the nearest tens and hundreds	3
4.N.27	Check reasonableness of an answer by using estimation	21
STRAND 2: ALGEBRA		
Variables and Expressions: Students will represent and analyze algebraically a wide variety of problem solving situations.		
4.A.1	Evaluate and express relationships using open sentences with one operation	22, 23
Equations and Inequalities: Students will perform algebraic procedures accurately.		
4.A.2	Use the symbols $<$, $>$, $=$, and $\neq$ (with and without the use of a number line) to compare whole numbers and unit fractions and decimals (up to hundredths)	2, 14, 17, 23
4.A.3	Find the value or values that will make an open sentence true, if it contains $<$ or $>$	23
Patterns, Relations and Functions: Students will recognize, use, and represent algebraically patterns, relations, and functions.		
4.A.4	Describe, extend, and make generalizations about numeric ($+$, $-$, $\times$, $\div$) and geometric patterns	24
4.A.5	Analyze a pattern or a whole-number function and state the rule, given a table or an input/output box	25

Indicator	New York State Grade 4 Math Indicators	Coach Lesson(s)
STRAND 3: GEOMETRY		
Shapes: Students will use visualization and spatial reasoning to analyze characteristics and properties of geometric shapes.		
4.G.1	Identify and name polygons, recognizing that their names are related to the number of sides and angles (triangle, quadrilateral, pentagon, hexagon, and octagon)	28
4.G.2	Identify points and line segments when drawing a plane figure	28
4.G.3	Find perimeter of polygons by adding sides	29
4.G.4	Find the area of a rectangle by counting the number of squares needed to cover the rectangle	29
4.G.5	Define and identify vertices, faces, and edges of three-dimensional shapes	30
Geometric Relationships: Students will identify and justify geometric relationships, formally and informally.		
**4.G.6	Draw and identify intersecting, perpendicular, and parallel lines	26
**4.G.7	Identify points and rays when drawing angles	27
**4.G.8	Classify angles as acute, obtuse, right, and straight	27
STRAND 4: MEASUREMENT		
Units of Measurement: Students will determine what can be measured and how, using appropriate methods and formulas.		
4.M.1	Select tools and units (customary and metric) appropriate for the length being measured	31, 32
4.M.2	Use a ruler to measure to the nearest standard unit (whole, $\frac{1}{2}$ and $\frac{1}{4}$ inches, whole feet, whole yards, whole centimeters, and whole meters)	31, 32
4.M.3	Know and understand equivalent standard units of length: 12 inches = 1 foot 3 feet = 1 yard	31
4.M.4	Select tools and units appropriate to the mass of the object being measured (grams and kilograms)	33
4.M.5	Measure mass, using grams	33
4.M.6	Select tools and units appropriate to the capacity being measured (milliliters and liters)	34
4.M.7	Measure capacity, using milliliters and liters	34
Units: Students will use units to give meaning to measurements.		
4.M.8	Make change, using combined coins and dollar amounts	35
4.M.9	Calculate elapsed time in hours and half hours, not crossing A.M./P.M.	36
4.M.10	Calculate elapsed time in days and weeks, using a calendar	36

* Grade 3 May–June Indicators   ** Grade 4 May–June Indicators

Indicator	New York State Grade 4 Math Indicators	Coach Lesson(s)
	STRAND 5: STATISTICS AND PROBABILITY	
Collection of Data: Students will collect, organize, display, and analyze data.		
*3.S.1	Formulate questions about themselves and their surroundings	37
*3.S.2	Collect data using observation and surveys, and record appropriately	37
**4.S.1	Design investigations to address a question from given data	39
**4.S.2	Collect data using observations, surveys, and experiments and record appropriately	39
Organization and Display of Data		
4.S.3	Represent data using tables, bar graphs, and pictographs	38
Analysis of Data		
4.S.4	Read and interpret line graphs	41
Predictions from Data: Students will make predictions that are based upon data analysis.		
4.S.5	Develop and make predictions that are based on data	40
4.S.6	Formulate conclusions and make predictions from graphs	40

			NYS Math Indicators
Lesson 1	Place Value	14	4.N.1, 4.N.2, 4.N.4, 4.N.5
Lesson 2	Compare and Order Whole Numbers	20	4.N.3, 4.A.2
Lesson 3	Round Whole Numbers	26	4.N.26
Lesson 4	Add and Subtract Whole Numbers	31	4.N.5, 4.N.14
Lesson 5	Multiplication Facts	37	4.N.16
Lesson 6	Properties of Multiplication	43	4.N.6, 4.N.13
Lesson 7	Multiply by One-Digit Whole Numbers	48	4.N.18
Lesson 8	Multiply by Two-Digit Whole Numbers	52	4.N.19
Lesson 9	Division Facts	56	4.N.16, 4.N.17
Lesson 10	Divide by One-Digit Divisors	62	4.N.21, 4.N.22
Lesson 11	Multiply and Divide by Multiples of 10 and 100	70	4.N.20
Lesson 12	Fractions	74	4.N.7
Lesson 13	Equivalent Fractions	77	3.N.14*, 4.N.8
Lesson 14	Compare and Order Fractions	84	3.N.15*, 4.N.9, 4.A.2
Lesson 15	Add and Subtract Fractions	91	4.N.23
Lesson 16	Decimals	96	4.N.10, 4.N.11
Lesson 17	Compare and Order Decimals	103	4.N.12, 4.A.2
Lesson 18	Add and Subtract Decimals	109	4.N.25
Lesson 19	Fraction-Decimal Equivalencies	115	4.N.24
Lesson 20	Solve Real-World Problems	121	4.N.15
Lesson 21	Use Estimation to Check Answers	126	4.N.27
Strand 1 Review		131	

* Grade 3 May–June Indicators

1 Place Value

4.N.1, 4.N.2, 4.N.4, 4.N.5

Getting the Idea

A **whole number** can be written in different forms:

standard form: 2,346

word form: two thousand three hundred forty-six

expanded form: 2,000 + 300 + 40 + 6

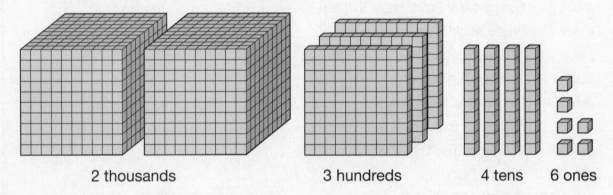

2 thousands   3 hundreds   4 tens   6 ones

You can use a **place-value chart** to write numbers. **Place value** is the value of a digit in a number based on its location.

Remember: 10 ones = 1 ten

10 tens = 1 hundred

10 hundreds = 1 thousand

10 thousands = 1 ten thousand

Ten Thousands	Thousands	Hundreds	Tens	Ones
1	0,	0	0	0

EXAMPLE 1

What are the next two numbers in this pattern?

4,200, 5,200, 6,200, 7,200, _____, _____

STRATEGY   **Identify the pattern. Then extend the pattern.**

STEP 1   Identify the pattern.

The pattern is skip count by 1,000s.

STEP 2   Extend the pattern.

Count forward by 1,000s.

8,200, 9,200

SOLUTION   **The next two numbers in the pattern are 8,200 and 9,200.**

EXAMPLE 2

How is 7,412 written in word form?

STRATEGY   **Use place value. Look at the comma.**

STEP 1   How many thousands in all?

There are 7 thousands.

Write *seven thousand*. Put a comma (,) after *thousand*.

STEP 2   How many hundreds in all?

There are 4 hundreds.

Write *seven thousand, four hundred*.

STEP 3   How many tens and ones in all?

There are 12 ones.

Write *seven thousand, four hundred twelve*.

SOLUTION   **The word form of 7,412 is *seven thousand, four hundred twelve*.**

EXAMPLE 3

What is the value of the digit 2 in 4,269?

STRATEGY   **Use a place-value chart.**

STEP 1   Write the number in a place-value chart.

Thousands	Hundreds	Tens	Ones
4,	2	6	9

STEP 2   Find the value of the digit 2.

The digit 2 is in the hundreds place.

The value of the digit is 12 × 100, or 200.

SOLUTION   **The value of the digit 2 is 2 hundreds, or 200.**

EXAMPLE 4

How do you write 9,875 in expanded form?

STRATEGY   **Use a place-value chart.**

STEP 1   Write the number in a place value chart.

Thousands	Hundreds	Tens	Ones
9,	8	7	5

STEP 2   Write the value of each digit.

9 thousands = 9,000

8 hundreds = 800

7 tens = 70

5 ones = 5

STEP 3   Write the expanded form.

9,000 + 800 + 70 + 5

SOLUTION   **The number 9,875 in expanded form is 9,000 + 800 + 70 + 5.**

EXAMPLE 5

Mr. Salvo's fourth-grade class collected cans for the annual recycling fair at school. They need to collect 300 cans to win a prize. How many tens are in 3 hundred?

STRATEGY   **Use place value to find how many tens.**

STEP 1   Review how many tens are in 1 hundred.

10 tens = 1 hundred

STEP 2   Find how many tens are in 3 hundred.

If 10 tens = 1 hundred, then 30 tens = 3 hundred.

SOLUTION   **There are 30 tens in 3 hundred.**

COACHED EXAMPLE

Write the standard form of this number:

six thousand three hundred fourteen

THINKING IT THROUGH

Start with the thousands.

How many thousands? _____

How many hundreds? _____

How many tens? _____

How many ones? _____

Write the digits from thousands to ones. _____

The standard form of the number six thousand three hundred fourteen is

_____ .

Lesson Practice

Choose the correct answer.

1. What is three thousand, one hundred nineteen in standard form?

   A. 3,119

   B. 3,190

   C. 3,191

   D. 3,911

2. Which digit is in the thousands place in 7,436?

   A. 3

   B. 4

   C. 6

   D. 7

3. What is the value of the digit 8 in 9,805?

   A. 8,000

   B. 800

   C. 80

   D. 8

4. Which number has the digit 6 in the thousands place and in the tens place?

   A. 3,562

   B. 4,668

   C. 6,361

   D. 6,629

5. In 2008, Hamilton County, New York had a population of 5,021 people. What is this number in expanded form?

   A. 50,000 + 2,000 + 100 + 10

   B. 5,000 + 2,000 + 100 + 1

   C. 5,000 + 200 + 1

   D. 5,000 + 20 + 1

6. Which is the same as 200 tens?

   A. 2 ones

   B. 20 ones

   C. 2 hundreds

   D. 20 hundreds

7. Keisha read a total of 3,480 pages of mystery novels during the summer. What is another way to write this number?

   A. three hundred forty-eight

   B. three thousand forty-eight

   C. three thousand four hundred eighty

   D. thirty thousand four hundred eighty

8. New York State entered the Union, becoming a state in the United States of America, on July 26, 1788. What is 1788 in expanded form?

   A. 1,000 + 700 + 80 + 8
   B. 1,000 + 700 + 80
   C. 1,000 + 700 + 8
   D. 1,000 + 70 + 8

9. Write the number 6,337 in word form.

   Answer Six thousand, three hundred thirty three.

10. What is the value of the digit 7 in 9,376?

    Answer 70

EXTENDED-RESPONSE QUESTION

11. Thomas collected 2,382 marbles in a jar.

    Part A Write the number of marbles in expanded form.

    2000 + 300 + 80 + 2

    Part B Write the number of marbles in word form.

    two thousand, three hundred eighty two

2 Compare and Order Whole Numbers

4.N.3, 4.A.2

Getting the Idea

Use place value to compare and order whole numbers. When comparing numbers, use these symbols.

The symbol > means **is greater than**.

The symbol < means **is less than**.

The symbol = means **is equal to**.

The symbol   means **is not equal to**.

EXAMPLE 1

Which symbol makes this statement true?

8,972 ◯ 8,959

STRATEGY   **Line up the numbers on the ones place. Remember to move from left to right. Start comparing the digits in the greatest place.**

STEP 1   Line up the numbers on the ones place.

8,972

8,959

STEP 2   Compare the digits in the thousands place.

Since 8 = 8, compare the next greatest place.

STEP 3   Compare the digits in the hundreds place.

Since 9 = 9, compare the next greatest place.

STEP 4   Compare the digits in the tens place.

Since 7 > 5, then 8,972 > 8,959.

SOLUTION   8,972 ⊙> 8,959

EXAMPLE 2

The students at Hanover Elementary School collected bottle caps to use for a school project. The fourth graders collected 5,654 caps and the fifth graders collected 5,469 caps.

Which number is greater: 5,654 or 5,469? Use > or < in your answer.

STRATEGY   **Use a number line to compare the two numbers.**

STEP 1   Draw a number line from 5,000 to 6,000.

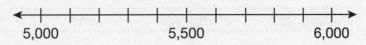

STEP 2   Show 5,654 and 5,469 on the number line.

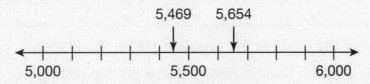

STEP 3   Compare the location of the numbers.

5,654 is to the right of 5,469 on the number line.

So, 5,654 is greater than 5,469.

SOLUTION   **5,654 > 5,469**

When you order three or more numbers, one number will be the greatest, and another will be the least.

EXAMPLE 3

Order the following numbers from least to greatest: 7,828; 7,625; 7,832.

STRATEGY   **Use place value to compare the numbers.**

STEP 1   Write the numbers in a place-value chart.

Thousands	Hundreds	Tens	Ones
7,	8	2	8
7,	6	2	5
7,	8	3	2

STEP 2   Compare the digits in the greatest place: thousands.

The digits in the thousands place are the same.

STEP 3   Compare the digits in the next greatest place: hundreds.

8 hundreds > 6 hundreds

7,828 > 7,625

7,832 > 7,625

The least number is 7,625.

STEP 4   For the remaining numbers, compare the digits in the third greatest place: tens.

3 tens > 2 tens

7,832 > 7,828

The greatest number is 7,832.

SOLUTION   **From least to greatest, the order of the numbers is: 7,625; 7,828; 7,832.**

COACHED EXAMPLE

Order the numbers below from greatest to least.

   2,519   2,486   2,603   2,549

THINKING IT THROUGH

Use place value to compare the numbers.

Write the numbers in a place-value chart.

Thousands	Hundreds	Tens	Ones

Compare the digits in the greatest place: _____.

All of the digits in the thousands place are the _____.

Compare the digits in the next greatest place: _____.

_____ hundreds > _____ hundreds > _____ hundreds

The greatest number is _____.

The least number _____.

To compare the remaining numbers, compare the digits in the next greatest place: _____.

_____ tens > _____ ten

So _____ > _____.

The numbers in order from greatest to least are

_____; _____; _____; _____.

Lesson Practice

Choose the correct answer.

1. Which symbol makes this statement true?

   8,452 ◯ 8,352

   A. >
   B. <
   C. =
   D.

2. Which sentence is true?

   A. 8,412 > 9,421
   B. 7,905 < 6,905
   C. 9,058 > 9,037
   D. 2,836 > 2,915

3. Which list shows the numbers in order from least to greatest?

   A. 7,358; 7,536; 7,185; 7,581
   B. 7,536; 7,358; 7,581; 7,185
   C. 7,581; 7,185; 7,536; 7,358
   D. 7,185; 7,358; 7,536; 7,581

4. Which digit makes this sentence true?

   6,485 < 6, ☐ 42

   A. 5
   B. 4
   C. 3
   D. 2

5. Which shows the members of the Stamp Club in order from the greatest number of stamps in their collection to the least?

   Stamp Collections

Stamp Club Member	Number of Stamps
Paula	1,583
Marcus	1,487
Leonard	1,569
Jackie	1,607
Frank	1,513

   A. Marcus, Frank, Leonard, Paula, Jackie
   B. Jackie, Paula, Leonard, Frank, Marcus
   C. Marcus, Leonard, Jackie, Paula, Frank
   D. Leonard, Jackie, Paula, Frank, Marcus

6. Which of the following is greater than 8,278 and less than 9,384?

   A. 8,209

   B. 8,728

   C. 9,394

   D. 9,438

7. Which of the following statements is **not** true?

   A. 6,329 < 6,801

   B. 1,698 > 1,408

   C. 6,496 7,221

   D. 2,190 = 2,668

8. Eddie collects coins. He has a penny from 1912, a nickel from 1915, a dime from 1917, and a quarter from 1921. Which of the coins is the oldest?

   Answer _____

9. Which number is the greatest: 3,564; 3,465; or 3,645?

   Answer _____

EXTENDED-RESPONSE QUESTION

10. The table below shows the number of ice cream cones sold at Bennie's Ice Cream Stand each month for four months.

Ice Cream Cones Sold

Month	Number of Cones
May	1,296
June	1,474
July	1,107
August	1,443

   Part A   In which month was the least number of ice cream cones sold?

   _____

   Part B   In which months was the number of ice cream cones sold greater than 1,300?

   _____

3 Round Whole Numbers

4.N.26

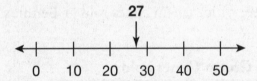

Getting the Idea

You can **round** a number to the nearest ten or hundred. Rounding gives a close number to use when exact numbers are not needed. It is often easier to compute with rounded numbers.

You can use a number line when rounding numbers. It can help you decide what number to round to.

EXAMPLE 1

What is 27 rounded to the nearest ten?

STRATEGY   Use a number line.

STEP 1   Place 27 on a number line.

27

↓

0   10   20   30   40   50

STEP 2   Decide whether 27 is closer to 20 or 30.

27 is closer to 30.

Round 27 up to 30.

SOLUTION   **27 rounded to the nearest ten is 30.**

You can also use these rules when rounding numbers.

> ### Rules for Rounding
>
> Look at the digit to the right of the place you are rounding to.
> - If the digit is 1, 2, 3, or 4, round down. Leave the digit in the rounding place as is.
> - If the digit is 5, 6, 7, 8, or 9, round up. Increase the digit in the rounding place by 1.
> - Change the digits to the right of the rounding place to zeros.

EXAMPLE 2

Carmen is driving from New York City to Buffalo, NY. The distance is 363 miles. What is this number rounded to the nearest ten?

STRATEGY   **Use the rounding rules to round to the nearest ten.**

    STEP 1   Underline the place you want to round to, the tens place.

         3<u>6</u>3

    STEP 2   Look at the digit to the right of the rounding place, the ones place.

         3<u>6</u>3

         The digit is 3. It is less than 5, so round down.

    STEP 3   The rounding place digit stays the same. The digit to the right of the rounding place changes to zero.

         363 rounds down to 360.

SOLUTION   **Rounded to the nearest ten, 363 is 360.**

EXAMPLE 3

Martin's Music Shop earned $445 in April, $625 in May, and $575 in June. During which two months did Martin's Music Shop earn about the same amount of money?

STRATEGY   **Round each amount to the nearest hundred. Then compare.**

STEP 1   Round to the nearest hundred.

The shop earned $445 in April.

The tens digit is 4, round down.

$445 → $400

The shop earned $625 in May.

The tens digit is 2, round down.

$625 → $600

The shop earned $575 in June.

The tens digit is 7, round up.

$575 → $600

STEP 2   Compare the rounded amounts.

April: $400

May: $600

June: $600

SOLUTION   **Martin's Music Shop earned about the same amount of money in May and June.**

COACHED EXAMPLE

There are 948 students in Lakeville Elementary School. To the nearest ten and hundred, about how many students are there at Lakeville Elementary School?

THINKING IT THROUGH

First round 948 to the nearest ten.

The digit in the tens place is _____.

The digit to the right of the rounding place is _____.

This digit is _____ than 5.

Since the digit to the right is greater than 5, round _____.

Change the digit to the right of the rounding place to _____.

Rounded to the nearest ten, 948 is _____.

Next round 948 to the nearest hundred.

The digit in the hundreds place is _____.

The digit to the right of the rounding place is _____.

This digit is _____ than 5.

Since the digit to the right is less than 5, round _____.

Change all the digits to the right of the rounding place to _____.

Rounded to the nearest hundred, 948 is _____.

Lesson Practice

Choose the correct answer.

1. What is 719 rounded to the nearest hundred?

    A. 700   C. 750

    B. 720   D. 800

2. Which number rounds to 400?

    A. 489   C. 352

    B. 451   D. 349

3. What is 92 rounded to the nearest ten?

    A. 100   C. 90

    B. 95   D. 80

4. Which number does **not** round to 50?

    A. 45   C. 52

    B. 48   D. 55

5. Which shows 126 rounded to the nearest ten?

    A. 100   C. 150

    B. 130   D. 200

6. Francesca has 281 marbles in her marble collection. About how many marbles does she have, to the nearest hundred?

    A. 300   C. 280

    B. 290   D. 200

7. Rod scored 24 goals over two seasons of soccer. About how many goals did he score, rounded to the nearest ten?

    A. 10   C. 30

    B. 20   D. 40

8. Which number rounds to 900?

    A. 839   C. 919

    B. 845   D. 951

9. Lotus baked 328 cookies for her class bake sale. What is 328 rounded to the nearest ten?

    Answer _____

10. Steven is thinking of a number between 40 and 50 that, when rounded to the nearest ten, is 50. What could be Steven's number?

    Answer _____

4 Add and Subtract Whole Numbers

4.N.5, 4.N.14

Getting the Idea

The numbers you **add** are **addends** and the answer is the **sum**.

When using paper and pencil, add the digits from right to left. If the sum of the digits in a column is 10 or greater, you will have to **regroup**.

EXAMPLE 1

Leo flew 2,582 miles from New York City to San Francisco. Then he flew 2,557 miles from San Francisco to Honolulu. How many miles did Leo fly in all to get to Honolulu?

STRATEGY Align the digits on the ones place. Add from right to left.

STEP 1 Add the ones: 2 + 7 = 9

$$\begin{array}{r} 2{,}58\mathbf{2} \\ +\ 2{,}55\mathbf{7} \\ \hline 9 \end{array}$$

STEP 2 Add the tens: 8 + 5 = 13

Write the 3.

Regroup the 1 ten.

$$\begin{array}{r} {}^{1} \\ 2{,}582 \\ +\ 2{,}557 \\ \hline 39 \end{array}$$

STEP 3 Add the hundreds: 5 + 5 + 1 = 11

Write the 1.

Regroup the 1 ten.

$$\begin{array}{r} {}^{1\ 1} \\ 2{,}582 \\ +\ 2{,}557 \\ \hline 139 \end{array}$$

STEP 4 Add the thousands: 2 + 2 + 1 = 5

$$\begin{array}{r} {}^{1\ 1} \\ 2{,}582 \\ +\ 2{,}557 \\ \hline \mathbf{5}{,}139 \end{array}$$

SOLUTION Leo flew 5,139 miles in all to get from New York City to Honolulu.

Here are the parts in a **subtraction** sentence.

$$3,667 \leftarrow \textbf{minuend}$$
$$- 1,243 \leftarrow \textbf{subtrahend}$$
$$\overline{2,424} \leftarrow \textbf{difference}$$

Subtract the digits from right to left. Sometimes you may need to regroup when subtracting.

EXAMPLE 2

$5,125 - 1,778 = \square$

STRATEGY   **Rewrite the problem. Subtract from right to left. Regroup when necessary.**

STEP 1   5 ones < 8 ones so regroup 1 ten as 10 ones.

Subtract the ones:
$15 - 8 = 7$

$$\begin{array}{r} {}^{1\ 15} \\ 5,1\,2\,5 \\ -\,1,7\,7\,8 \\ \hline 7 \end{array}$$

STEP 2   1 ten < 7 tens so regroup 1 hundred as 10 tens.

Subtract the tens:
$11 - 7 = 4$

$$\begin{array}{r} {}^{11} \\ {}^{0\ 1\ 15} \\ 5,1\,2\,5 \\ -\,1,7\,7\,8 \\ \hline 4\,7 \end{array}$$

STEP 3   0 hundreds < 7 hundreds so regroup 1 thousand as 10 hundreds.

Subtract the hundreds:
$10 - 7 = 3$

$$\begin{array}{r} {}^{10\ 11} \\ 4\ 0\ 1\ 15 \\ 5,1\,2\,5 \\ -\,1,7\,7\,8 \\ \hline 3\,4\,7 \end{array}$$

STEP 4   Subtract the thousands:
$4 - 1 = 3$

$$\begin{array}{r} {}^{10\ 11} \\ 4\ 0\ 1\ 15 \\ 5,1\,2\,5 \\ -\,1,7\,7\,8 \\ \hline 3,3\,4\,7 \end{array}$$

SOLUTION $5,125 - 1,778 = 3,347$

EXAMPLE 3

$4,007 - 1,526 = \square$

STRATEGY   **Rewrite the problem. Subtract from right to left. Regroup when necessary.**

STEP 1   Subtract the ones: $7 - 6 = 1$.

$$
\begin{array}{r}
4,0\,0\,\mathbf{7} \\
-\,1,5\,2\,\mathbf{6} \\
\hline
\mathbf{1}
\end{array}
$$

STEP 2   0 tens < 2 tens, so regroup 1 thousand as 10 hundreds. Then regroup 1 hundred as 10 tens. Subtract the tens: $10 - 2 = 8$

$$
\begin{array}{r}
9 \\
3 \;\; \cancel{10} \;\; 10 \\
\cancel{4},\cancel{0}\,0\,7 \\
-\,1,5\,\mathbf{2}\,\mathbf{6} \\
\hline
\mathbf{8}\,1
\end{array}
$$

STEP 3   Subtract the hundreds: $9 - 5 = 4$

Then subtract the thousands: $3 - 1 = 2$

$$
\begin{array}{r}
9 \\
3 \;\; \cancel{10} \;\; 10 \\
\cancel{4},\cancel{0}\,\cancel{0}\,7 \\
-\,1,5\,2\,6 \\
\hline
2,\mathbf{4}\,8\,1
\end{array}
$$

SOLUTION $4,007 - 1,526 = 2,481$

You can check the answer to a subtraction problem by adding the difference and the subtrahend.

$$
\begin{array}{r}
4,0\,0\,7 \\
-\,1,5\,2\,6 \\
\hline
2,4\,8\,1
\end{array}
\qquad
\begin{array}{r}
^{1}\;\;^{1} \\
2,4\,8\,1 \\
+\,1,5\,2\,6 \\
\hline
4,0\,0\,7
\end{array}
$$

The sum is 4,007, so the difference is correct.

COACHED EXAMPLE

According to the 2000 census, East Greenbush, New York had a population of 4,085 and Mount Sinai, New York had a population of 8,734. How much greater was the population of Mount Sinai than the population of East Greenbush in 2000?

THINKING IT THROUGH

Align the digits on the ones place. Subtract from right to left.

$$
\begin{array}{r}
8,734 \\
-\ 4,085 \\
\hline
\end{array}
$$

Do you need to regroup the ones? _____

Do you need to regroup the tens? _____

Do you need to regroup the hundreds? _____

Do you need to regroup the thousands? _____

Subtract from right to left.

Use addition to check your answer.

Is your answer correct?_____

The population of Mount Sinai was _____ more than the population of East Greenbush in 2000.

Lesson Practice

Choose the correct answer.

1. Add.

   3,674
   + 4,369

   A. 7,933

   B. 7,943

   C. 8,033

   D. 8,043

2. Subtract.

   8,715
   − 5,923

   A. 2,792

   B. 2,802

   C. 2,892

   D. 3,792

3. Sara scored 4,293 points on a computer game. Her score was 1,536 points more than Elvin's score on the same game. How many points did Elvin score?

   A. 3,757

   B. 2,767

   C. 2,757

   D. 2,667

4. In 2007, there were 2,034 country radio stations, 557 sports radio stations, and 711 oldies radio stations in the United States. How many more country stations were there than sports and oldies stations combined?

   A. 766

   B. 856

   C. 1,268

   D. 3,302

5. What is the sum of 4,215 and 3,107?

   A. 7,302

   B. 7,312

   C. 7,321

   D. 7,322

6. What is the difference of 7,038 − 1,843?

   A. 5,195

   B. 5,295

   C. 6,195

   D. 6,295

7. Mr. Peters flew 1,224 miles in July and 857 miles in August. How many miles did he fly in all during the two months?

   A. 2,071

   B. 2,077

   C. 2,081

   D. 2,087

8. What is the difference of the following subtraction problem?

   $$9,715$$
   $$- 7,923$$

   A. 1,792

   B. 2,792

   C. 2,882

   D. 2,892

9. At the local tire factory, 3,166 tires were made on Thursday and 2,941 tires were made on Friday. How many tires were made on Thursday and Friday in all?

   Answer _____ tires

10. William I of Great Britain began his reign in 1066. William IV, the last William to be king, began his reign in 1830. How many years passed between the beginnings of the reigns of William I and William IV?

    Answer _____ years

5 Multiplication Facts

 4.N.16

Getting the Idea

When you **multiply**, you combine equal groups. It is a shortcut for repeated addition.

The numbers you multiply are called **factors** and the answer is the **product**. In $3 \times 2 = 6$, 3 and 2 are the factors and 6 is the product.

3	×	2	=	6
↑		↑		↑
factor		factor		product

One way to show multiplication is to use an **array**. An array has the same number of objects in each row.

EXAMPLE 1

What multiplication fact does this array show?

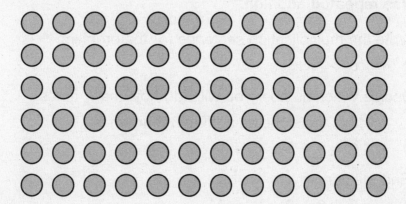

STRATEGY Count the number of rows. Then count the number of counters in each row.

STEP 1   Count the number of rows.

There are 6 rows of counters.

STEP 2   Count the number of counters in each row.

There are 12 counters in each row.

STEP 3   Count the number of counters.

There are 72 counters in all.

SOLUTION   **The array shows 6 × 12 = 72.**

You can also use repeated addition to solve a multiplication problem.
Repeated addition is adding the same number over and over again.

Suppose you didn't know the answer to 3×2. You could do two things.

Add 3 two times: $3 + 3 = 6$.

Add 2 three times: $2 + 2 + 2 = 6$.

EXAMPLE 2

Stephanie baked 4 pies for her school's bake sale. She cut each pie into 8 slices.
How many slices of pie did Stephanie make in all?

STRATEGY   **Use repeated addition.**

STEP 1   Write the multiplication sentence for the problem.

$4 \times 8 = \square$

STEP 2   Decide which number you will add how many times.

4×8 is the same as adding 8 four times.

$4 \times 8 = 8 + 8 + 8 + 8$

STEP 3   Find the sum.

$8 + 8 + 8 + 8 = 32$

SOLUTION   **Stephanie made 32 slices of pie in all.**

You can use a multiplication table to help you learn the basic facts of multiplication.

To use a multiplication table, find the box where the row and column of the two factors meet.

Columns

×	0	1	2	3	4	5	6	7	8	9	10	11	12
0	0	0	0	0	0	0	0	0	0	0	0	0	0
1	0	1	2	3	4	5	6	7	8	9	10	11	12
2	0	2	4	6	8	10	12	14	16	18	20	22	24
3	0	3	6	9	12	15	18	21	24	27	30	33	36
4	0	4	8	12	16	20	24	28	32	36	40	44	48
5	0	5	10	15	20	25	30	35	40	45	50	55	60
6	0	6	12	18	24	30	36	42	48	54	60	66	72
7	0	7	14	21	28	35	42	49	56	63	70	77	84
8	0	8	16	24	32	40	48	56	64	72	80	88	96
9	0	9	18	27	36	45	54	63	72	81	90	99	108
10	0	10	20	30	40	50	60	70	80	90	100	110	120
11	0	11	22	33	44	55	66	77	88	99	110	121	132
12	0	12	24	36	48	60	72	84	96	108	120	132	144

Rows

EXAMPLE 3

Multiply 9 × 5.

STRATEGY   **Use the multiplication table.**

STEP 1   Look at the 9s column.

STEP 2   Find the 5s row.

STEP 3   Find the number inside the box where the 9s column and the 5s row meet. That is the product.

SOLUTION   **9 × 5 = 45**

COACHED EXAMPLE

Kate gives her dog 3 biscuits each day. How many biscuits does Kate give her dog in 7 days?

THINKING IT THROUGH

Use the multiplication table.

The total number of biscuits is _____ × _____.

Find the column in the table for _____.

Find the row in the table for _____.

The number in the box where this column and row meet is _____.

3 × 7 = _____

Kate gives her dog _____ biscuits in 7 days.

Lesson Practice

Choose the correct answer.

1. What is 7×6?

   A. 35

   B. 36

   C. 42

   D. 49

2. For a park cleanup, volunteers were assigned to teams of 6. There were 4 teams. How many volunteers helped clean the park?

   A. 24

   B. 28

   C. 32

   D. 36

3. What multiplication fact is shown by this repeated addition?

   $$3 + 3 + 3 + 3 + 3 + 3$$

   A. $5 \times 3 = 15$

   B. $6 \times 3 = 18$

   C. $7 \times 3 = 21$

   D. $8 \times 3 = 24$

4. A dime is worth 10 cents. How much are 7 dimes worth?

   A. 7 cents

   B. 35 cents

   C. 70 cents

   D. 100 cents

5. Which of the following has the greatest product?

   A. $9 \times 6 = 30$

   B. $10 \times 5 \ 50$

   C. $11 \times 4 \ 44$

   D. 12×3

6. What multiplication fact does this array show?

   A. $2 \times 10 = 20$

   B. $2 \times 5 = 10$

   C. $5 \times 5 = 25$

   D. $2 \times 6 = 12$

7. Millie baked 5 trays of muffins. Each tray holds 6 muffins. Which expression could be used to show how many muffins Millie baked in all?

A. 5 + 6 + 5 + 6

B. 5 + 5 + 5 + 5 + 5

C. 6 + 6 + 6 + 6 + 6

D. 6 + 6 + 6 + 6

8. Herschel runs 4 miles each day. How many miles does he run in a week? (1 week = 7 days)

*Answer*_____ miles

9. Each debate team has 3 students. A tournament will have 12 teams. How many students will participate in the tournament?

*Answer*_____ students

EXTENDED-RESPONSE QUESTION

10. There are 12 rows of seats on a school bus. Each row can fit 4 students.

Part A Write a multiplication sentence that shows how many students can sit on the bus.

_____

Part B How many students can sit on the bus?

_____

6 Properties of Multiplication

4.N.6, 4.N.13

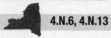

Getting the Idea

An **even number** is a number that has a 0, 2, 4, 6, or 8 in the ones place.

An **odd number** is a number that has a 1, 3, 5, 7, or 9 in the ones place.

The table below shows the products of multiplying even factors and odd factors.

Factor × Factor = Product	Example
even number × even number = even number	$2 \times 2 = 4$
odd number × odd number = odd number	$3 \times 3 = 9$
odd number × even number = even number	$3 \times 2 = 6$

EXAMPLE 1

Which of the following numbers, when multiplied by an odd number, will have an even product?

A. 3   **B.** 5   **C.** 8   **D.** 9

STRATEGY   **Multiply each number by an odd factor. Use the table to help you decide which will have an even product.**

STEP 1   Check choice A: 3.

3 is an odd number.

Remember that odd × even = even.

3 × odd number = odd number

STEP 2   Check choice B: 5.

5 is an odd number.

5 × odd number = odd number

STEP 3   Check choice C: 8.

8 is an even number.

Remember that even × odd = even

8 × odd number = even number

STEP 4   Check choice D: 9.

9 is an odd number.

9 × odd number = odd number

SOLUTION   **The answer is choice C. The number 8, when multiplied by an odd number, will have an even product.**

The **associative property of multiplication** states that the grouping of the factors does not change a product.

$(7 \times 5) \times 8 = 7 \times (5 \times 8)$

$35 \times 8 = 7 \times 40$

$280 = 280$

EXAMPLE 2

$(12 \times 3) \times 4 = \square$

STRATEGY   **Use the associative property of multiplication to find the product.**

STEP 1   Use the associative property to regroup the factors.

$(12 \times 3) \times 4 = 12 \times (3 \times 4)$

STEP 2   Multiply inside the parentheses first.

$12 \times (3 \times 4) = 12 \times 12$

STEP 3   Find the product.

$12 \times 12 = 144$

SOLUTION   **$(12 \times 3) \times 4 = 144$**

COACHED EXAMPLE

A school bus has 11 rows of seats. There are 4 seats in each row. How many seats are there on 2 buses?

THINKING IT THROUGH

Write an expression for the problem:

(_____ rows × _____ seats in each row) × _____ buses

Use the associative property of multiplication to regroup the factors.

11 × (_____ × _____)

First, multiply inside the parentheses.

(4 × 2) = _____

Then find the product.

11 × _____ = _____

There are _____ seats on 2 buses.

Lesson Practice

Choose the correct answer.

1. Which number goes in the box to make the number sentence true?

    $7 \times \square$ = odd number

    A. 4

    B. 5

    C. 6

    D. 8

2. Which shows the associative property of multiplication?

    A. $8 \times 0 = 0$

    B. $7 \times 3 \times 1 = 7 \times 3$

    C. $(6 \times 8) \times 5 = 6 \times (8 \times 5)$

    D. $5 \times 12 = 12 \times 5$

3. Which of the following statements is **not** true?

    A. 6×4 = odd number

    B. 7×3 = odd number

    C. 2×5 = even number

    D. 4×4 = even number

4. Which number when multiplied by any odd number always results in an odd number?

    A. 2

    B. 4

    C. 7

    D. 8

5. Which is another way to write the expression $27 \times (15 \times 19)$?

    A. $(27 + 15) \times 19$

    B. $27 \times (15 + 19)$

    C. $27 + (15 + 19)$

    D. $(27 \times 15) \times 19$

6. Francine wrote a number sentence to show that an even number multiplied by another even number results in an even product. Which could be Francine's number sentence?

    A. $12 \times 16 = 192$

    B. $20 \times 13 = 260$

    C. $15 \times 23 = 345$

    D. $31 \times 14 = 434$

7. Which expression is equivalent to $(9 \times 6) \times 10$?

   A. $(9 + 6) \times 10$

   B. $9 \times (6 + 10)$

   C. 15×10

   D. 54×10

8. Which number sentence shows that an odd number multiplied by an even number equals an even number?

   A. $3 \times 5 = 15$

   B. $5 \times 6 = 30$

   C. $7 \times 9 = 63$

   D. $8 \times 6 = 48$

9. Ted wrote the number sentence $9 \times (5 \times 3) = 135$. Connor rewrote Ted's number sentence by grouping the numbers differently. Connor told Ted that he used the associative property of multiplication. Write Connor's number sentence.

   *Answer*_____

10. Lilly and Anita worked on their math homework together. Lilly wrote that $5 \times 6 =$ odd number. Anita told Lilly that this was wrong. She reminded Lilly about the properties of even and odd numbers. What property did Lilly forget?

    *Answer*_____

EXTENDED-RESPONSE QUESTION

11. The music room has 12 rows of seats on 2 sides of the piano. Each row of seats has 5 chairs. The music teacher wrote the following number sentence to represent how many seat there are in all.

    $(12 \times 2) \times 5 = \square$

    Part A   Use the associative property to write this number sentence another way.

    _____

    Part B   Solve your number sentence to find how many seats there are in all.

    _____

7 Multiply by One-Digit Whole Numbers

4.N.18

Getting the Idea

You can multiply a two-digit number by a one-digit number by using basic facts and regrouping.

Sometimes using models can help you multiply.

EXAMPLE 1

Find 24 × 3.

STRATEGY Use models to multiply.

STEP 1 Use models to show 3 groups of 24.

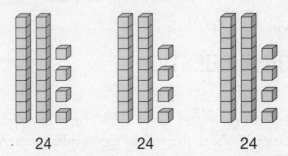

24   24   24

STEP 2 Combine the tens models. Then combine the ones models.

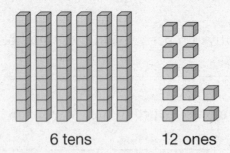

6 tens   12 ones

STEP 3   Regroup 12 ones as 1 ten and 2 ones.

Add the tens and ones.

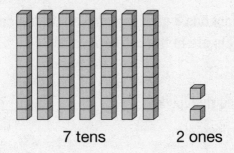

7 tens   2 ones

SOLUTION $24 \times 3 = 72$

EXAMPLE 2

$67 \times 5 = \square$

STRATEGY   **Multiply the ones. Then multiply the tens.**

STEP 1   Write the problem in vertical form.

Multiply the ones: $7 \times 5 = 35$

Write the 5 and regroup the 3 tens.

$$\begin{array}{r} {\scriptstyle 3} \\ 6\,7 \\ \times\ \ 5 \\ \hline 5 \end{array}$$

STEP 2   Multiply the tens: $6 \times 5 = 30$

Add the regrouped tens: $30 + 3 = 33$

Write the 33.

$$\begin{array}{r} {\scriptstyle 3} \\ 6\,7 \\ \times\ \ 5 \\ \hline 3\,3\,5 \end{array}$$

SOLUTION $67 \times 5 = 335$

COACHED EXAMPLE

Empire Elementary School has 4 sections of seats in the cafeteria. Each section has 48 seats. How many seats are in the cafeteria in all?

THINKING IT THROUGH

What numbers should you multiply? _____ × _____

Write the problem in vertical form.

Multiply the ones: 4 × 8 = _____

Write the 2 in the answer and regroup _____ tens.

Multiply the tens: 4 × 4 = _____

Add the _____ regrouped tens. Write the_____ in the answer.

There are _____ seats in the cafeteria in all.

Lesson Practice

Choose the correct answer.

1. $78 \times 4 = \square$

   A. 282

   B. 292

   C. 302

   D. 312

2. A tour group of 68 people has an overnight stay at a motel in Buffalo. Each person will receive a 3-pancake breakfast. How many pancakes will the motel serve to the tour group?

   A. 184   C. 204

   B. 194   D. 214

3. What is 43×8?

   A. 324   C. 344

   B. 342   D. 354

4. What is 37×9?

   A. 323   C. 343

   B. 333   D. 433

5. The drive from Albany to Poughkeepsie, New York, is 76 miles. Mr. Carter drives that route round trip 3 times each month. How many miles does he drive each month? (Hint: Round trip means there and back.)

   A. 456   C. 356

   B. 426   D. 228

6. Which does **not** have a product of 288?

   A. 57×5 C. 48×6

   B. 72×4 D. 96×3

7. Mr. Garcia has a cabinet with 8 shelves. There are 52 CDs on each shelf. How many CDs does Mr. Garcia have in his cabinet?

   Answer_____

8. There are 96 students in the fourth grade at Winfrey Elementary School. Each of the students took 5 tests this week. How many tests have to be graded by the fourth-grade teachers?

   Answer_____

Lesson

8 Multiply by Two-Digit Whole Numbers

4.N.19

Getting the Idea

To find the product of 2 two-digit factors, first multiply a factor by the ones digit of the other factor. Then multiply the same factor by the tens digit of the other factor. Add the partial products to find the product.

EXAMPLE 1

What is 45 × 28?

STRATEGY   **Multiply by each place value, regrouping when necessary.**

STEP 1   Multiply by the ones: 8 × 45.

Regroup as necessary.

$$
\begin{array}{r}
{}^{4} \\
4\,5 \\
\times\,2\,8 \\
\hline
3\,6\,0
\end{array}
$$
 ← partial product of 8 × 45

STEP 2   Multiply by the tens: 2 tens × 45.

Regroup as necessary.

Write a 0 in the ones place because you are multiplying the tens.

$$
\begin{array}{r}
{}^{1} \\
4\,5 \\
\times\,2\,8 \\
\hline
3\,6\,0 \\
9\,0\,0
\end{array}
$$
 ← partial product of 20 × 45

STEP 3   Add the partial products to find the product.

$$
\begin{array}{r}
4\,5 \\
\times\,2\,8 \\
\hline
3\,6\,0 \\
+\,9\,0\,0 \\
\hline
1{,}2\,6\,0
\end{array}
$$

SOLUTION   45 × 28 = 1,260

EXAMPLE 2

Multiply: 59 × 27

STRATEGY   **Multiply, regrouping when necessary.**

STEP 1   Multiply the 7 by the first factor, 59. Regroup as necessary.

$$\begin{array}{r} {}^{6} \\ 5\,9 \\ \times\,2\,7 \\ \hline 4\,1\,3 \end{array}$$ ← partial product of 7 × 59

STEP 2   Multiply the 2 tens by the first factor, 59.

$$\begin{array}{r} {}^{1} \\ \cancel{6} \\ 5\,9 \\ \times\,2\,7 \\ \hline 4\,1\,3 \\ 1\,1\,8\,0 \end{array}$$ ← partial product of 20 × 59

STEP 3   Add the two partial products to find the product.

$$\begin{array}{r} 1 \\ \cancel{6} \\ 5\,9 \\ \times\,2\,7 \\ \hline 4\,1\,3 \\ +\,1\,1\,8\,0 \\ \hline 1{,}5\,9\,3 \end{array}$$

SOLUTION   **The product of 59 × 27 is 1,593.**

COACHED EXAMPLE

Find the product.

$$
\begin{array}{r}
2\,6 \\
\times\ 3\,4 \\
\hline
\end{array}
$$

THINKING IT THROUGH

What is the first step? Multiply _____ by _____.

What is the product of 4 × 26? _____

Remember to put a _____ in the ones place before multiplying the tens.

What is the next step? Multiply _____ tens by _____.

What is the product of 3 tens × 26? _____

What is the sum of the two partial products?

_____ + _____ = _____

The product of 26 × 34 is _____.

Lesson Practice

Choose the correct answer.

1. $37 \times 37 = \square$

   A. 1,369

   B. 1,381

   C. 1,401

   D. 1,419

2. $23 \times 21 = \square$

   A. 231

   B. 243

   C. 443

   D. 483

3. Find the product.

   $$\begin{array}{r} 96 \\ \times\ 62 \\ \hline \end{array}$$

   A. 5,580

   B. 5,642

   C. 5,952

   D. 6,052

4. Multiply: 74×11

   A.   81

   B.   747

   C.   814

   D. 7,474

5. Multiply: 24×15

   A. 200

   B. 260

   C. 300

   D. 360

6. Find the product.

   $$\begin{array}{r} 38 \\ \times\ 14 \\ \hline \end{array}$$

   A. 532

   B. 502

   C. 432

   D. 380

7. What is the product of 55 multiplied by 60?

   Answer _____

8. What is the product of 32 multiplied by 64?

   Answer _____

9 Division Facts

4.N.16, 4.N.17

Getting the Idea

When you **divide**, you separate into equal groups. In division, the number that is divided is the **dividend**. The number that divides the dividend is the **divisor**. The answer to a division problem is the **quotient**.

63	÷	9	=	7
↑		↑		↑
total number		number of groups		number in each group
dividend		divisor		quotient

EXAMPLE 1

Michael divides 15 marbles into 3 equal groups. How many marbles are in each group?

STRATEGY   Make a model to find the quotient.

STEP 1   Put out 15 counters.

STEP 2   Divide the counters into 3 equal rows.

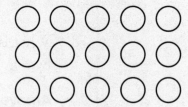

STEP 3   Find the number of counters in each row.

There are 5 counters in each row.

SOLUTION $15 \div 3 = 5$

You can also use repeated subtraction to solve a division problem. Repeated subtraction is subtracting the same number over and over again.

EXAMPLE 2

Sandra bakes 24 cookies that she shares equally with 3 of her friends. How many cookies does Sandra give each friend?

STRATEGY   **Use repeated subtraction.**

STEP 1   How many cookies are there in all?

There are 24 cookies in all.

STEP 2   How many people are sharing the cookies?

Sandra and her 3 friends are sharing the cookies together.

There are 4 people sharing the cookies.

STEP 3   Subtract 4 from 24 until you get 0.

$24 - 4 = 20$

$20 - 4 = 16$

$16 - 4 = 12$

$12 - 4 = 8$

$8 - 4 = 4$

$4 - 4 = 0$

STEP 4   How many times was 4 subtracted from 24?

Four was subtracted from 24 six times.

$24 \div 4 = 6$

SOLUTION   **Sandra gives each friend 6 cookies.**

Division is the opposite, or inverse, of multiplication. This means division undoes multiplication. You can use **fact families** to learn multiplication and division facts.

A fact family is a group of related facts that use the same numbers. The fact family for 7, 9, and 63 is: $7 \times 9 = 63$, $9 \times 7 = 63$, $63 \div 7 = 9$, and $63 \div 9 = 7$.

A multiplication table can help you write fact families.

×	0	1	2	3	4	5	6	7	8	9	10	11	12
0	0	0	0	0	0	0	0	0	0	0	0	0	0
1	0	1	2	3	4	5	6	7	8	9	10	11	12
2	0	2	4	6	8	10	12	14	16	18	20	22	24
3	0	3	6	9	12	15	18	21	24	27	30	33	36
4	0	4	8	12	16	20	24	28	32	36	40	44	48
5	0	5	10	15	20	25	30	35	40	45	50	55	60
6	0	6	12	18	24	30	36	42	48	54	60	66	72
7	0	7	14	21	28	35	42	49	56	63	70	77	84
8	0	8	16	24	32	40	48	56	64	72	80	88	96
9	0	9	18	27	36	45	54	63	72	81	90	99	108
10	0	10	20	30	40	50	60	70	80	90	100	110	120
11	0	11	22	33	44	55	66	77	88	99	110	121	132
12	0	12	24	36	48	60	72	84	96	108	120	132	144

EXAMPLE 3

Use the numbers 10, 12, and 120 to write a fact family.

STRATEGY **Write the multiplication and division facts that use these numbers.**

   STEP 1   Write two multiplication facts.

$$10 \times 12 = 120$$
$$12 \times 10 = 120$$

   STEP 2   Write two division facts.

$$120 \div 10 = 12$$
$$120 \div 12 = 10$$

SOLUTION **The fact family is $10 \times 12 = 120$, $12 \times 10 = 120$, $120 \div 10 = 12$, and $120 \div 12 = 10$.**

COACHED EXAMPLE

Ms. Lopez divided the 35 desks in her classroom into 5 equal rows. How many desks are in each row?

THINKING IT THROUGH

You can write this problem as _____ ÷ _____.

Use a multiplication fact you know.

5 × _____ = 35

Use the inverse operation to find the quotient.

If 5 × 7 = 35, then 35 ÷ 5 = _____.

There are _____ desks in each row.

Lesson Practice

Choose the correct answer

1. What is $49 \div 7$?

   A. 6

   B. 7

   C. 8

   D. 9

2. Last week, Marsha wrote 28 pages in her journal. She wrote the same number of pages each day. How many pages did she write each day? (1 week = 7 days)

   A. 3

   B. 4

   C. 5

   D. 6

3. Which fact does **not** belong in the same fact family as the others?

   A. $8 \times 2 = 16$

   B. $16 \div 2 = 8$

   C. $4 \times 4 = 16$

   D. $16 \div 8 = 2$

4. Which of the following has a quotient of 4?

   A. $20 \div 5$

   B. $30 \div 6$

   C. $32 \div 4$

   D. $36 \div 12$

5. Vanessa used repeated subtraction to divide 96 by 8. How many times did she subtract 8 from 96?

   A. 9

   B. 10

   C. 11

   D. 12

6. Which division fact is related to $6 \times 8 = 48$?

   A. $56 \div 8 = 7$

   B. $48 \div 8 = 6$

   C. $36 \div 6 = 6$

   D. $42 \div 6 = 7$

7. If $4 \times 3 = 12$, which number belongs in the box?

   $$12 \div 3 = \square$$

   A. 2

   B. 3

   C. 4

   D. 12

8. Lawrence wants to check that he solved the division problem below correctly.

$$54 \div 9 = 6$$

Which number sentence could Lawrence use to check his answer?

A. $54 \times 6 = \square$

B. $9 \times 54 = \square$

C. $9 \div 6 = \square$

D. $9 \times 6 = \square$

9. Mr. Frankel divides his 36 paintbrushes into 3 equal groups for a class art project. How many paintbrushes does Mr. Frankel put in each group?

Answer_____

10. Lee solves the following division problem.

$$60 \div 5 = 12$$

Write a number sentence that Lee could use to check her answer.

Answer _____

EXTENDED-RESPONSE QUESTION

11. Gary brought 24 mini-muffins to the breakfast meeting. There are 6 people at the meeting, including Gary.

Part A How many muffins will each person get if each person at the meeting is given the same number of muffins? Show your work.

Part B Write a number sentence to check your answer.

_____

10 Divide by One-Digit Divisors

 4.N.21, 4.N.22

Getting the Idea

Division problems can be written these ways.

$$\text{divisor} \overline{)\text{dividend}}^{\text{quotient}} \qquad \text{dividend} \div \text{divisor} = \text{quotient}$$

To divide by a 1-digit divisor, look at the greatest place in the dividend. If the divisor is less than or equal to that digit, divide. Otherwise, divide by the first two places of the dividend.

EXAMPLE 1

What is 56 ÷ 4?

STRATEGY   Use the inverse operation of multiplication.

STEP 1   Make a model using 56 counters. Make 4 equal rows.

STEP 2   Count the number of counters in each row.

There are 14 counters in each of the 4 rows.

$4 \times 14 = 56$

SOLUTION $56 \div 4 = 14$

EXAMPLE 2

Alex has 72 baseball cards that he wants to divide equally into three boxes. How many baseball cards will he put in each box?

STRATEGY   **Divide the dividend by the divisor at each place from left to right.**

> STEP 1   Divide 72 into 3 equal groups.
>
> Divide the tens by 3.
>
> There are 2 tens. Multiply and then subtract.
>
> $$\begin{array}{r} 2 \\ 3\overline{)72} \\ -6 \\ \hline 1 \end{array}$$ ← Multiply: 3×2 tens = 6 tens
> ← Subtract: 7 tens − 6 tens = 1 ten
>
> STEP 2   Bring down the ones and divide.
>
> There are 4 ones. Multiply and then subtract.
>
> $$\begin{array}{r} 24 \\ 3\overline{)72} \\ -6 \\ \hline 12 \\ -12 \\ \hline 0 \end{array}$$ ← Multiply: 3×4 ones = 12 ones
> ← Subtract: 12 ones − 12 ones = 0 ones

SOLUTION   **There will be 24 baseball cards in each box.**

A **remainder** is a number that is left after division has been completed. A remainder must be less than a divisor. For example, 10 ÷ 3 = 3 R1.

EXAMPLE 3

What is the quotient for 61 ÷ 7?

STRATEGY   **Divide the dividend by the divisor in each place, from left to right.**

STEP 1   Divide the tens by 7.

There are not enough tens, so divide the ones.

$$
\begin{array}{r}
8 \\
7\overline{)61} \\
-56 \\
\hline
5
\end{array}
$$

← Multiply: 7 × 8 = 56
← Subtract: 61 − 56 = 5

STEP 2   Find how many are left over.

There are 5 left over.

This is the remainder.

STEP 3   Write the remainder.

$$
\begin{array}{r}
8 \quad \textbf{R5} \\
7\overline{)61} \\
-56 \\
\hline
5
\end{array}
$$

SOLUTION   **61 ÷ 7 = 8 R5**

You can check your answer by multiplying the divisor by the quotient and then adding the remainder. For Example 3, multiply 7 × 8 = 56. Then add the remainder: 56 + 5 = 61. 61 is the dividend, so your quotient is correct.

After you have solved the problem, you may need to interpret the remainder. There are three ways to interpret a remainder depending on the problem you are solving.

1. Drop the remainder.

2. The remainder is the answer.

3. Add 1 to the quotient.

EXAMPLE 4

There are 38 chorus members going to a concert. Each parent driver can take 4 students. How many parent drivers are needed to get all the chorus members to the concert?

STRATEGY   **Divide. Then interpret the remainder.**

    STEP 1   Divide $38 \div 4$.

                $38 \div 4 = 9$ R2

    STEP 2   Interpret the remainder.

                Since the 2 students remaining need to be driven, one more car is needed.

                Add 1 to the quotient.

SOLUTION   **To get all the chorus members to the concert, 10 parent drivers are needed.**

COACHED EXAMPLE

The fourth-grade class collected 45 empty bottles for a recycling project. The students are packing them into boxes that hold 6 bottles each. The recycling depot accepts only full boxes. How many boxes will the fourth grade class take to the recycling depot?

THINKING IT THROUGH

Divide: _____ bottles ÷ _____ bottles in a box.

There are _____ full boxes.

There are _____ bottles left over.

Interpret the remainder.

Since the _____ remaining bottles cannot fill a box, _____ the remainder.

The fourth grade will take _____ boxes to the recycling depot.

Lesson Practice

Choose the correct answer.

1. What is $57 \div 3$?

   A. 18

   B. 19

   C. 20

   D. 21

2. Mr. Telmond has 84 CDs in 3 cases. Each case contains an equal number of CDs. How many CDs are in each case?

   A. 38

   B. 36

   C. 30

   D. 28

3. Jess has a 96-page book to read over spring break. She wants to read the same number of pages each day for 6 days. How many pages should she read each day?

   A. 15

   B. 16

   C. 17

   D. 18

4. What is $65 \div 3$?

   A. 20 R4

   B. 21 R1

   C. 21 R3

   D. 21 R2

5. Juanita borrowed $98 to buy a computer game system. She will make 7 equal payments to pay it off. How much money will Juanita pay each time?

   A. $16

   B. $15

   C. $14

   D. $13

6. The total cost for 3 tickets to the game is $81. What is the cost for each ticket?

   A. $27

   B. $28

   C. $37

   D. $38

7. There are 76 players trying out for youth basketball teams. Each team will have exactly 9 players. How can you interpret the remainder?

   A. Drop the remainder.

   B. Add 1 to the quotient.

   C. The remainder is the answer.

   D. Subtract 1 from the quotient.

8. What is the quotient?

   $$59 \div 5$$

   A. 12 R4

   B. 12 R2

   C. 11 R4

   D. 11 R2

9. $72 \div 4 = \square$

   A. 17

   B. 18

   C. 27

   D. 28

10. Which division sentence has a remainder?

    A. $99 \div 3$

    B. $38 \div 2$

    C. $77 \div 6$

    D. $65 \div 5$

11. Nicole found that $85 \div 4 = 21$ R1. Which can you use to check if Nicole's answer is correct?

    A. $21 \times 4 = \square$, then add 1

    B. $85 \times 21 = \square$, then subtract 1

    C. $85 \times 4 = \square$, then subtract 1

    D. $21 \times 85 = \square$, then add 1

12. $48 \div 3 = \square$

    A. 26

    B. 25

    C. 16

    D. 15

13. Erik solved the equation $85 \div 5 = 17$. Write a number sentence that Erik could use to check his answer.

    Answer _____

14. There are a total of 69 ladybugs in 3 bughouses in the school science lab. Each bughouse has the same number of ladybugs. How many ladybugs are there in each bughouse?

    Answer _____

EXTENDED-RESPONSE QUESTION

15. Mr. Santos needs enough pizzas for 91 fourth-grade students. Each pizza has 8 slices. How many pizzas does Mr. Santos need so that each student can have 1 slice?

Part A Write the number sentence for this problem.

_____

Part B Solve the number sentence from Part A. Explain how you interpreted the remainder.

_____

_____

_____

11 Multiply and Divide by Multiples of 10 and 100

4.N.20

Getting the Idea

To multiply any number by 10, place a zero at the end of the number. This is because 10 has one zero in it.

$$10 \times 8 = 80 \qquad 10 \times 76 = 760$$

To multiply any number by 100, put two zeros at the end of the number. This is because 100 has two zeros in it.

$$100 \times 8 = 800 \qquad 100 \times 76 = 7,600$$

EXAMPLE 1

What is 53×10?

STRATEGY   **10 has one zero, so place one zero at the end of the number.**

Any number multiplied by 10 is that number with one zero at the end of it.

$53 \times 10 = 530$

SOLUTION   **$53 \times 10 = 530$**

A **multiple** of a number is the result of multiplying the number by a counting number (1, 2, 3, 4, 5...).

A multiple of 10 is any counting number multiplied by 10.

A multiple of 100 is any counting number multiplied by 100.

Multiples of 10 end with one zero (10, 20, 30, 40, 50, ...).

Multiples of 100 end with two zeros (100, 200, 300, 400, 500, ...).

EXAMPLE 2

Jackie rides her bicycle 13 miles a day. If she does this for 100 days, how many miles will she ride in all?

STRATEGY   **Multiply. Place zeros at the end of the factor.**

STEP 1   Write a multiplication sentence.

$13 \times 100 = \underline{\hspace{1cm}}$

STEP 2   Multiply.

Any number multiplied by 100 is that number with two zeros at the end of the number.

$13 \times 100 = 1,300$

SOLUTION   **Jackie will ride 1,300 miles in all.**

Dividing a multiple of 10 by 10 is the reverse of multiplying by 10. Instead of putting a zero at the end of the number, you take away a zero. The same is true for dividing multiples of 100 by 100. Here are some examples.

$370 \div 10 = 37$ $1,400 \div 100 = 14$

EXAMPLE 3

Tanya has collected 70 dimes in a jar. She is going to package her dimes in paper rolls of 10 to bring to the bank. How many rolls of dimes can she make?

STRATEGY   **Divide by taking away a zero.**

STEP 1   Write a number sentence.

$70 \div 10 = \underline{\hspace{1cm}}$

STEP 2   Divide.

Since 70 is a multiple of 10, when you divide it by 10, you take away one zero.

$70 \div 10 = 7$

SOLUTION   **Tanya can make 7 rolls of dimes.**

COACHED EXAMPLE

The high school gymnasium bleachers hold 900 people. The bleachers are separated into sections that each hold 100 people. How many sections of bleachers are there in the high school gymnasium?

THINKING IT THROUGH

Decide how to solve the problem.

Write a number sentence. _____ ÷ _____

Is 900 a multiple of 100? _____

Divide by taking away _____ zeros.

_____ ÷ _____ = _____

There are _____ sections of bleachers in the high school gymnasium.

Lesson Practice

Choose the correct answer.

1. Multiply.

    $10 \times 9 = \square$

    A. 9

    B. 90

    C. 109

    D. 900

2. Multiply.

    $100 \times 7 = \square$

    A. 700

    B. 170

    C. 70

    D. 7

3. Divide.

    $1,000 \div 100 = \square$

    A. 1,000

    B. 100

    C. 10

    D. 1

4. Divide.

    $1,300 \div 100 = \square$

    A. 1

    B. 13

    C. 30

    D. 130

5. Samuel is buying shirts for the soccer team. The shirts cost $8 each. He orders a box of 100 shirts. How much does he pay in all?

    Answer $_____

6. Annie writes that 60×10 is 6. Annie knows that her answer is wrong but cannot remember what to do. How many zeros should Annie write after the 6 to solve 60×10?

    Answer _____

12 Fractions

 4.N.7

Getting the Idea

A **fraction** names part of a whole or part of a group. The **numerator** is the top number. It tells how many equal parts are being considered. The **denominator** is the bottom number. It tells how many equal parts there are in all.

You can locate a fraction on a number line. The denominator represents the number of equal parts there are between whole numbers.

EXAMPLE 1

What fraction does the letter R represent on the number line?

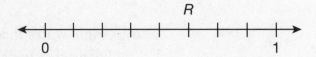

STRATEGY   **Find the number of equal parts.**

STEP 1   How many equal parts are between 0 and 1?

There are 8 spaces between 0 and 1 on the number line.

That means there are 8 equal parts. This will represent the denominator.

STEP 2   Which mark is R located on?

It is on the 5th mark after zero. This will represent the numerator.

STEP 3   Write the fraction.

Since the numerator is 5 and the denominator is 8, the fraction is written as $\frac{5}{8}$.

SOLUTION   The letter R represents the fraction $\frac{5}{8}$.

A fraction bar also represents division. So $\frac{1}{2}$ is the same as $1 \div 2$.

EXAMPLE 2

What is another way to write $\frac{3}{4}$?

STRATEGY   **Rewrite the fraction as a division problem.**

The fraction bar means division.

Divide the numerator by the denominator.

$\frac{3}{4} = 3 \div 4$

SOLUTION   **Another way to write $\frac{3}{4}$ is 3 ÷ 4.**

COACHED EXAMPLE

What fraction does the letter *S* represent on the number line?

THINKING IT THROUGH

Find the number of equal parts.

There are _____ equal parts between 0 and 1.

The denominator will be _____.

S is located on the _____ mark after zero.

The numerator will be _____.

Write the fraction.

The fraction is _____.

The letter *S* represents the fraction _____ on the number line.

Lesson Practice

Choose the correct answer.

1. Where is the letter *A* located on the number line?

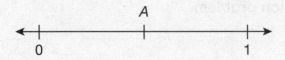

   A. $\frac{1}{2}$

   B. $\frac{2}{4}$

   C. $\frac{2}{6}$

   D. $\frac{4}{6}$

2. Where is the letter *B* located on the number line?

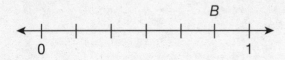

   A. $\frac{2}{6}$

   B. $\frac{3}{6}$

   C. $\frac{4}{6}$

   D. $\frac{5}{6}$

3. Where is the letter *C* located on the number line?

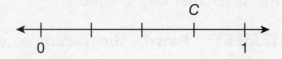

   A. $\frac{1}{4}$

   B. $\frac{2}{4}$

   C. $\frac{3}{4}$

   D. $\frac{4}{5}$

4. Which is another way to represent $\frac{2}{3}$?

   A. $2 \div 3$

   B. $3 \div 2$

   C. 2×3

   D. 3×2

5. Which is another way to represent $\frac{5}{12}$?

   A. 5×12

   B. 12×5

   C. $5 \div 12$

   D. $12 \div 5$

6. Write another way to represent $\frac{3}{10}$.

   Answer_____

13 Equivalent Fractions

3.N.14, 4.N.8

Getting the Idea

Equivalent fractions are fractions that use different numerators and denominators but have the same value.

EXAMPLE 1

Write two equivalent fractions that name the shaded parts of the circle.

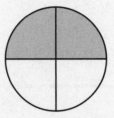

STRATEGY **Look at the shaded parts of the circle in two ways.**

STEP 1   Count the number of equal parts. Count the number of shaded parts.

There are 4 equal parts. There are 2 shaded parts.

So $\frac{2}{4}$ of the circle is shaded.

STEP 2   Look at the shaded parts another way.

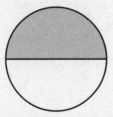

The two shaded parts are half of the circle.

One half of the circle is shaded.

So $\frac{1}{2}$ of the circle is shaded.

SOLUTION   **The fractions $\frac{1}{2}$ and $\frac{2}{4}$ are equivalent fractions that both name the shaded parts of the circle.**

You can use fraction strips to find equivalent fractions.

EXAMPLE 2

What number makes this sentence true?

$$\frac{2}{3} = \frac{\Box}{6}$$

$\frac{1}{3}$	$\frac{1}{3}$	$\frac{1}{3}$

$\frac{1}{6}$	$\frac{1}{6}$	$\frac{1}{6}$	$\frac{1}{6}$	$\frac{1}{6}$	$\frac{1}{6}$

STRATEGY   **Use fraction strips.**

STEP 1   Use two $\frac{1}{3}$ fraction strips to show $\frac{2}{3}$.

STEP 2   Use enough $\frac{1}{6}$ fraction strips to make the same length as $\frac{2}{3}$.

It takes four $\frac{1}{6}$ fraction strips to equal two $\frac{1}{3}$ fraction strips.

Four $\frac{1}{6}$ fraction strips equals $\frac{4}{6}$.

SOLUTION $\frac{2}{3}$ $\frac{4}{6}$

You can use number lines to find equivalent fractions.

EXAMPLE 3

What fraction with a denominator of 5 is equal to $\frac{6}{10}$?

STRATEGY   **Use a number line to find an equivalent fraction.**

STEP 1   Use a number line in tenths to locate $\frac{6}{10}$. Then draw a number line in fifths below the number line in tenths.

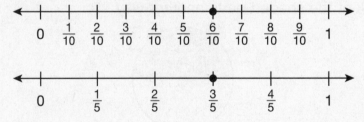

STEP 2   Write the fraction with a denominator of 5 that lines up with $\frac{6}{10}$.

$\frac{3}{5}$ lines up with $\frac{6}{10}$.

SOLUTION $\frac{6}{10}$ $\frac{3}{5}$

You can multiply or divide the numerator and denominator by the same number to find equivalent fractions.

EXAMPLE 4

Use multiplication to find an equivalent fraction for $\frac{3}{4}$.

STRATEGY **Multiply the numerator and denominator by the same number.**

STEP 1 Multiply both the numerator and denominator by 2.

$$\frac{3}{4} \times \frac{2}{2} = \frac{3 \times 2}{4 \times 2} = \frac{6}{8}$$

STEP 2 Check your work using fraction strips.

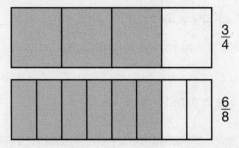

$\frac{3}{4}$ and $\frac{6}{8}$ have the same length.

SOLUTION $\frac{3}{4}$ and $\frac{6}{8}$ **are equivalent fractions.**

COACHED EXAMPLE

What fraction with a denominator of 12 is equivalent to $\frac{3}{6}$?

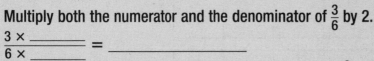

THINKING IT THROUGH

What is the denominator of $\frac{3}{6}$? _____

What will the denominator of the equivalent fraction be? _____

What number can you multiply 6 by to get a product of 12? _____

Multiply both the numerator and the denominator of $\frac{3}{6}$ by 2.

$$\frac{3 \times \rule{1.5cm}{0.4pt}}{6 \times \rule{1.5cm}{0.4pt}} = \rule{3cm}{0.4pt}$$

What equivalent fraction is created when you multiply $\frac{3}{6}$ by $\frac{2}{2}$? _____

_____ is the fraction with a denominator of 12 that is equivalent to $\frac{3}{6}$.

Lesson Practice

Choose the correct answer.

1. A fractional part of this group of squares is shaded.

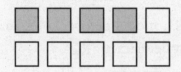

   Which group of triangles is shaded to represent a fraction with an equivalent value?

   A. △ △ △ △

   B. △ △ △ △ △

   C. △ △ △ △ △

   D. △ △ △ △ △ △

2. Which fraction is equivalent to $\frac{4}{6}$?

   A. $\frac{1}{4}$

   B. $\frac{1}{3}$

   C. $\frac{1}{2}$

   D. $\frac{2}{3}$

3. The fraction strip shows $\frac{10}{12}$.

   Which fraction is equivalent to $\frac{10}{12}$?

   A. $\frac{2}{3}$

   B. $\frac{3}{4}$

   C. $\frac{4}{5}$

   D. $\frac{5}{6}$

4. Which number makes this sentence true?

   $$\frac{1}{2} = \frac{\square}{4}$$

   A. 2

   B. 3

   C. 4

   D. 6

5. Which fraction is **not** equivalent to $\frac{1}{2}$?

   A. $\frac{2}{4}$

   B. $\frac{4}{6}$

   C. $\frac{5}{10}$

   D. $\frac{6}{12}$

6. Which fraction is equivalent to $\frac{3}{9}$?

   A. $\frac{1}{4}$

   B. $\frac{1}{3}$

   C. $\frac{1}{2}$

   D. $\frac{2}{3}$

7. What could you do to find an equivalent fraction for $\frac{6}{8}$?

   A. Divide the numerator and the denominator by 2.

   B. Divide the numerator by 2.

   C. Divide the denominator by 2.

   D. Multiply the numerator by 3.

8. Maria drew this rectangle.

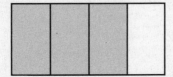

Which fraction is equivalent to the shaded part of Maria's rectangle?

   A. $\frac{3}{6}$

   B. $\frac{7}{8}$

   C. $\frac{6}{10}$

   D. $\frac{9}{12}$

9. The model is shaded to represent a fraction.

Which model below shows an equivalent fraction?

   A.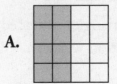

   B.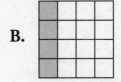

   C.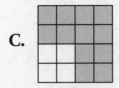

   D.

10. Which fraction is **not** equivalent to the fraction shown on the number line?

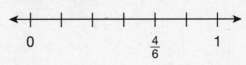

   A. $\frac{2}{3}$ C. $\frac{8}{12}$

   B. $\frac{3}{6}$ D. $\frac{12}{18}$

11. Name two fractions that are equivalent to $\frac{4}{10}$.

Answer_____

12. What number will make this sentence true?

$$\frac{2}{10} = \frac{8}{\square}$$

Answer_____

EXTENDED-RESPONSE QUESTION

13. Below are four figures divided into equal parts.

 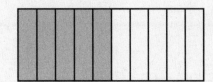 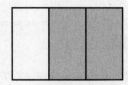

Figure A   **Figure B**   **Figure C**   **Figure D**

Part A   Which two figures are shaded to show equivalent fractions?

_____

Part B   Look at your answer to Part A. Draw a figure that shows another equivalent fraction to these two fractions.

14 Compare and Order Fractions

 3.N.15, 4.N.9, 4.A.2

Getting the Idea

You can use models to compare fractions. To compare **unit fractions**, which are fractions with a numerator of 1, look at the denominators. The fraction with the lesser denominator is the greater fraction. For example, $\frac{1}{4} > \frac{1}{5}$.

EXAMPLE 1

Which is the greater fraction, $\frac{1}{3}$ or $\frac{1}{4}$?

STRATEGY Make a drawing to show each fraction.

STEP 1 Draw rectangles to represent each fraction.

One rectangle has 3 equal parts. 1 part is shaded.

Another rectangle has 4 equal parts. 1 part is shaded.

STEP 2 Compare the shaded parts.

The shaded area that shows $\frac{1}{3}$ is bigger than the shaded area that shows $\frac{1}{4}$.

SOLUTION The greater fraction is $\frac{1}{3}$.

To compare fractions with the same denominator, look at the numerators. The greater fraction has the greater numerator. For example, $\frac{3}{5} > \frac{2}{5}$.

You can also use number lines to compare and order fractions. The number farther to the right on a number line is the greater number.

EXAMPLE 2

Compare $\frac{1}{5}$ and $\frac{3}{5}$ using a number line. Which is greater?

STRATEGY   **Determine which fraction is farther to the right on the number line.**

STEP 1   Determine how many total equal parts there are between whole numbers on the number line.

The denominator for both $\frac{1}{5}$ and $\frac{3}{5}$ is 5.

STEP 2   Draw a number line showing 5 equal parts.

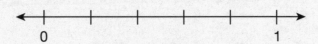

STEP 3   Write the fractions on the number line.

Each part represents $\frac{1}{5}$, so put a dot for $\frac{1}{5}$ and another for $\frac{3}{5}$.

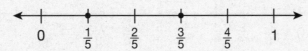

STEP 4   Compare the fractions.

Since $\frac{3}{5}$ is farther to the right, it is greater than $\frac{1}{5}$.

SOLUTION $\frac{3}{5}$ **is greater than** $\frac{1}{5}$**, or** $\frac{3}{5} > \frac{1}{5}$**.**

EXAMPLE 3

Compare the fractions $\frac{6}{8}$ and $\frac{3}{8}$. Which is greater?

STRATEGY   **Use fraction strips or models.**

STEP 1   Show $\frac{6}{8}$ and $\frac{3}{8}$ using fraction strips.

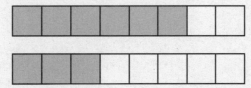

STEP 2   Compare the shaded areas.

The shaded area for $\frac{6}{8}$ is greater than the shaded area for $\frac{3}{8}$.

SOLUTION $\frac{6}{8}$ **is greater than** $\frac{3}{8}$**, or** $\frac{6}{8} > \frac{3}{8}$**.**

EXAMPLE 4

Order $\frac{7}{10}$, $\frac{3}{10}$, and $\frac{9}{10}$ from least to greatest.

STRATEGY   **Use a number line.**

STEP 1   Decide if the denominators are the same.

All of the denominators are 10.

STEP 2   Draw a number line showing 10 equal parts.

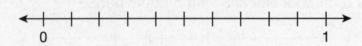

STEP 3   Write the fractions on the number line.

Each part represents $\frac{1}{10}$.

Put dots for $\frac{7}{10}$, $\frac{3}{10}$, and $\frac{9}{10}$.

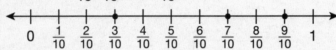

STEP 4   Order the fractions.

Since $\frac{9}{10}$ is farthest to the right, it is the greatest fraction.

Since $\frac{3}{10}$ is farthest to the left, it is the least fraction.

$\frac{7}{10}$ is in between $\frac{3}{10}$ and $\frac{9}{10}$.

So $\frac{3}{10} < \frac{7}{10} < \frac{9}{10}$

SOLUTION   **The fractions ordered from least to greatest are $\frac{3}{10}$, $\frac{7}{10}$, $\frac{9}{10}$.**

COACHED EXAMPLE

Order the following fractions from least to greatest.

$\frac{1}{2}$, $\frac{1}{8}$, $\frac{1}{4}$

THINKING IT THROUGH

Use fraction strips to _____ the fractions.

Shade the fraction strips for each fraction.

$\frac{1}{2}$

$\frac{1}{8}$

$\frac{1}{4}$

_____ has the greatest amount shaded.

_____ has the least amount shaded.

_____ is in between $\frac{1}{2}$ and $\frac{1}{8}$.

Compare the fractions.

_____ < _____ < _____

The fractions in order from least to greatest are _____, _____, _____.

Lesson Practice

Choose the correct answer.

1. $\frac{4}{5} \bigcirc \frac{2}{5}$

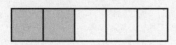

   A. <

   B. >

   C. =

   D. +

2. Which sentence is true?

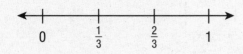

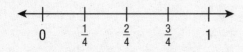

   A. $\frac{1}{3} < \frac{1}{4}$

   B. $\frac{1}{10} > \frac{1}{4}$

   C. $\frac{1}{4} < \frac{1}{10}$

   D. $\frac{1}{3} > \frac{1}{10}$

3. Which statement is correct?

   A. $\frac{1}{6} > \frac{3}{6}$

   B. $\frac{1}{6} < \frac{1}{6}$

   C. $\frac{1}{6} = \frac{3}{6}$

   D. $\frac{1}{6} < \frac{3}{6}$

4. Which statement is **not** correct?

   A. $\frac{7}{8} > \frac{5}{8}$

   B. $\frac{7}{8} > \frac{3}{8}$

   C. $\frac{5}{8} < \frac{1}{8}$

   D. $\frac{5}{8} < \frac{7}{8}$

5. Which of the following is ordered from least to greatest?

   A. $\frac{4}{5}, \frac{3}{5}, \frac{2}{5}$

   B. $\frac{2}{5}, \frac{3}{5}, \frac{4}{5}$

   C. $\frac{2}{5}, \frac{4}{5}, \frac{3}{5}$

   D. $\frac{3}{5}, \frac{4}{5}, \frac{2}{5}$

6. Which of the following is ordered from greatest to least?

A. $\frac{7}{10}, \frac{3}{10}, \frac{1}{10}$

B. $\frac{1}{10}, \frac{3}{10}, \frac{7}{10}$

C. $\frac{7}{10}, \frac{1}{10}, \frac{3}{10}$

D. $\frac{1}{10}, \frac{7}{10}, \frac{3}{10}$

7. Which statement about comparing fractions with the same denominator is true?

A. The less the numerator, the greater the fraction.

B. The less the numerator, the less the fraction.

C. The greater the numerator, the less the fraction.

D. The greater the denominator, the greater the fraction.

8. Which statement about comparing unit fractions is true?

A. The less the denominator, the greater the fraction.

B. The less the denominator, the less the fraction.

C. The greater the numerator, the less the fraction.

D. The greater the denominator, the greater the fraction.

9. Order $\frac{1}{5}, \frac{1}{8}$, and $\frac{1}{3}$ from least to greatest.

Answer _____ < _____ < _____

10. Write $\frac{3}{4}, \frac{1}{4}$, and $\frac{2}{4}$ in order from greatest to least.

Answer _____

EXTENDED-RESPONSE QUESTION

11. Marjorie needs $\frac{1}{5}$ cup of flour, $\frac{1}{4}$ cup of baking soda, and $\frac{1}{3}$ cup of sugar for her recipe.

    Part A   Which ingredient does she need the most of for her recipe? Which ingredient does she need the least of?

    Part B   Which part of the fraction did you need to look at to determine the answer to part A, the numerator or the denominator? Explain your answer.

    _____

    _____

    _____

15 Add and Subtract Fractions

4.N.23

Getting the Idea

Like fractions are fractions with the same denominator, which is called their **common denominator**.

When adding or subtracting like fractions, add or subtract the numerators and put the sum or difference over the common denominator.

EXAMPLE 1

Find the sum.

$$\frac{4}{6} + \frac{1}{6}$$

STRATEGY Add the fractions.

STEP 1 Do the fractions have a common denominator?

Yes, both fractions have a denominator of 6.

STEP 2 Add the numerators.

$$4 + 1 = 5$$

STEP 3 The denominator stays the same.

$$\frac{4}{6} + \frac{1}{6} = \frac{5}{6}$$

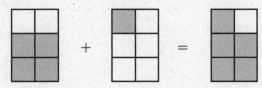

SOLUTION $\frac{4}{6} + \frac{1}{6} = \frac{5}{6}$

EXAMPLE 2

Allison bought a pizza divided into 10 slices. Allison's friends ate $\frac{7}{10}$ of the pizza and then her brother ate $\frac{2}{10}$ of the pizza. What fraction of the pizza was eaten in all?

STRATEGY Use fraction strips to model the problem.

STEP 1 Show $\frac{7}{10}$ with fraction pieces.

| $\frac{1}{10}$ | $\frac{1}{10}$ | $\frac{1}{10}$ | $\frac{1}{10}$ | $\frac{1}{10}$ | $\frac{1}{10}$ | $\frac{1}{10}$ | | | |

STEP 2 Add on $\frac{2}{10}$ with fraction pieces.

| $\frac{1}{10}$ | $\frac{1}{10}$ | $\frac{1}{10}$ | $\frac{1}{10}$ | $\frac{1}{10}$ | $\frac{1}{10}$ | $\frac{1}{10}$ | $\frac{1}{10}$ | $\frac{1}{10}$ | |

STEP 3 The fractions have a common denominator.

Count the numbers of $\frac{1}{10}$ fraction pieces.

There are nine $\frac{1}{10}$ fraction pieces in all.

| $\frac{1}{10}$ | $\frac{1}{10}$ | $\frac{1}{10}$ | $\frac{1}{10}$ | $\frac{1}{10}$ | $\frac{1}{10}$ | $\frac{1}{10}$ | $\frac{1}{10}$ | $\frac{1}{10}$ | |

$$\frac{1}{10}, \frac{2}{10}, \frac{3}{10}, \frac{4}{10}, \frac{5}{10}, \frac{6}{10}, \frac{7}{10}, \frac{8}{10}, \frac{9}{10}$$

STEP 4 Write the sum.

The numerator is 9. The denominator stays the same.

$$\frac{7}{10} + \frac{2}{10} = \frac{9}{10}$$

SOLUTION Allison's friends and brother ate $\frac{9}{10}$ of the pizza in all.

EXAMPLE 3

Find the difference. $\frac{6}{7} - \frac{4}{7}$

STRATEGY   **Subtract the fractions.**

    STEP 1   Do the fractions have a common denominator?

                   Yes, both fractions have a denominator of 7.

    STEP 2   Subtract the numerators.

$$6 - 4 = 2$$

    STEP 3   The denominator stays the same.

$$\frac{6}{7} - \frac{4}{7} = \frac{2}{7}$$

SOLUTION $\frac{6}{7} - \frac{4}{7} = \frac{2}{7}$

COACHED EXAMPLE

Sam walked $\frac{7}{8}$ mile to school. Toni walked $\frac{2}{8}$ mile to Patti's house. Then Toni and Patti walked $\frac{3}{8}$ mile to school. How much farther did Sam walk than Toni?

THINKING IT THROUGH

First, _____ to find how far Toni walked.

To add fractions with common denominators, add only the _____.

Add: _____ + _____ = _____

The denominator is _____.

Toni walked _____ mile.

Next, _____ to find how much farther Sam walked than Toni.

To subtract fractions with common denominators, subtract only the

_____.

Subtract: _____ − _____ = _____

The denominator is _____.

Sam walked _____ mile farther than Toni.

Lesson Practice

Choose the correct answer.

1. $\frac{3}{10} + \frac{5}{10} = \square$

   A. $\frac{2}{10}$

   B. $\frac{6}{10}$

   C. $\frac{8}{10}$

   D. $\frac{10}{10}$

2. Yesterday, Manny spent $\frac{1}{4}$ hour studying for his math quiz and $\frac{2}{4}$ hour writing a report. How long did Manny spend on schoolwork yesterday?

   A. $\frac{1}{4}$ hour

   B. $\frac{2}{4}$ hour

   C. $\frac{3}{4}$ hour

   D. 1 hour

3. Mary Ellen drew a rectangle that is $\frac{11}{12}$ inch long. The width of the rectangle is $\frac{5}{12}$ inch shorter. What is the width of Mary Ellen's rectangle?

   A. $\frac{4}{12}$ inch

   B. $\frac{6}{12}$ inch

   C. $\frac{7}{12}$ inch

   D. $\frac{9}{12}$ inch

4. $\frac{8}{10} - \frac{2}{10} = \square$

   A. $\frac{4}{10}$

   B. $\frac{5}{10}$

   C. $\frac{6}{10}$

   D. $\frac{8}{10}$

5. $\frac{3}{14} + \frac{7}{14} = \square$

   A. $\frac{10}{14}$

   B. $\frac{12}{14}$

   C. $\frac{13}{14}$

   D. $\frac{14}{14}$

6. Natalie ate $\frac{3}{8}$ of an apple. Later, she ate another $\frac{3}{8}$. What fraction of the apple did she eat in all?

   A. $\frac{2}{8}$

   B. $\frac{6}{8}$

   C. $\frac{7}{8}$

   D. $\frac{8}{8}$

7. Jenny read $\frac{2}{5}$ of her book on Saturday and $\frac{2}{5}$ of her book on Sunday. How much of her book did Jenny read over the weekend?

   A. $\frac{2}{5}$

   B. $\frac{3}{5}$

   C. $\frac{4}{5}$

   D. $\frac{5}{5}$

8. What is $\frac{3}{7} + \frac{2}{7}$?

   A. $\frac{2}{7}$

   B. $\frac{3}{7}$

   C. $\frac{4}{7}$

   D. $\frac{5}{7}$

9. In Perry's class, $\frac{5}{16}$ of the students have hazel eyes and $\frac{2}{16}$ of the students have blue eyes. What fraction of Perry's class has hazel or blue eyes?

   *Answer*_____

10. There was $\frac{3}{4}$ of a pad of paper on the counter. Greta used $\frac{1}{4}$ of the pad of paper to write thank-you letters. What fraction of the pad of paper is left?

    *Answer*_____

16 Decimals

4.N.10, 4.N.11

Getting the Idea

A **decimal** can name part of a whole or part of a group. A decimal can have a whole number part and a decimal part. A **decimal point (.)** separates the whole number part from the decimal part.

Hundreds	Tens	Ones		Tenths	Hundredths
9	4	3	.	6	5

The models below show one whole, one tenth, and one hundredth.

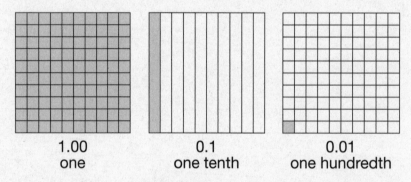

1.00
one

0.1
one tenth

0.01
one hundredth

To read and write a decimal less than 1, read the number to the right of the decimal point as you would read a whole number. Then read the least place value.

EXAMPLE 1

What decimal describes the part of the grid that is shaded? How would you read that decimal?

STRATEGY   Count the number of shaded squares. The entire grid equals 1.

STEP 1   There are 57 shaded squares.

There are 100 squares in the grid.

Fifty-seven hundredths of the grid is shaded.

The decimal is 0.57.

STEP 2   Determine how the decimal would be read.

57 decimals are shaded. The least place value is hundredths.

SOLUTION   **0.57 of the grid is shaded. The decimal is read as *fifty-seven hundredths*.**

To read and write a decimal greater than 1, read the whole number, the word *and*, then the part after the decimal point.

EXAMPLE 2

What is 18.24 in word form?

STRATEGY   **Use a place value chart.**

STEP 1   Write the number in the place value chart.

Tens	Ones		Tenths	Hundredths
1	8	.	2	4

STEP 2   Read the number from left to right.

Read the part to the left of the decimal point.

Write the whole part as *eighteen*.

Write the decimal point as *and*.

STEP 3   Read the part to the right of the decimal point.

The least place value of a decimal is the last word when writing a decimal.

Write the decimal part as *twenty-four hundredths*.

SOLUTION   **In word form, 18.24 is eighteen and twenty-four hundredths.**

To write a decimal in expanded form, write the decimal as the sum of each of its place values. For example, 18.24 is 10 + 8 + 0.2 + 0.04.

EXAMPLE 3

Write 63.79 in expanded form.

STRATEGY   **Use a place-value chart.**

STEP 1   Make the place-value chart.

Tens	Ones		Tenths	Hundredths
6	3	.	7	9

STEP 2   Write the value of each digit.

6 tens = 60

3 ones = 3

7 tenths = 0.7

9 hundredths = 0.09

STEP 3   Write the expanded form.

60 + 3 + 0.7 + 0.09

SOLUTION   **In expanded form, 63.79 is 60 + 3 + 0.7 + 0.09.**

Decimals can be used to represent money. The only difference between money and decimals is that in money you must remember to use the dollar sign.

1 dollar has the same value as 100 cents.

1 cent = $\frac{1}{100}$ of a dollar or $0.01

10 cents = $\frac{10}{100}$ of a dollar or $0.10

100 cents = $\frac{100}{100}$ of a dollar, one whole dollar, or $1.00

EXAMPLE 4

Dean has two dollars and sixty-seven cents in his pocket.

Write this amount in standard form.

STRATEGY **Write the place values from left to right.**

STEP 1   How many dollars?
There are 2 dollars.

STEP 2   How many cents?
There are 67 cents.

STEP 3   Write the amount as a decimal.
Remember to include the dollar sign.
2 dollars and 67 cents → $2.67

SOLUTION **In standard form, two dollars and sixty-seven cents is $2.67.**

COACHED EXAMPLE

Write the decimal represented by the grids in standard and expanded forms.

THINKING IT THROUGH

First write the decimal in standard form.

How many grids are completely shaded? _____

What whole number names the completely shaded grids? _____

How many squares are shaded in the partially shaded grid? _____

What decimal names the partially shaded grid? _____

In standard form, the decimal is _____.

Then write the decimal in expanded form.

How many ones? _____

How many tenths? _____

How many hundredths? _____

In expanded form, the decimal is _____.

Lesson Practice

Choose the correct answer.

1. Which decimal represents the shaded part of the grid?

    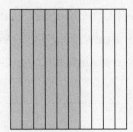

    A. 0.04

    B. 0.06

    C. 0.4

    D. 0.6

2. Which decimal represents the shaded part of the grid?

    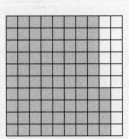

    A. 0.16

    B. 0.84

    C. 0.86

    D. 0.94

3. What is $400 + 90 + 0.7 + 0.05$ in standard form?

    A. 49.75

    B. 409.75

    C. 490.75

    D. 497.05

4. Which is another way to write five dollars and fifty-five cents?

    A. $5.05

    B. $5.15

    C. $5.50

    D. $5.55

5. What is three and nine hundredths written in standard form?

    A. 0.309   **C.** 3.09

    B. 0.39   **D.** 3.9

6. Wanda wrote a check for thirty-eight dollars and ninety-six cents. How is this written as a decimal dollar amount?

    A. $38.69

    B. $38.96

    C. $39.06

    D. $96.38

7. In the decimal 42.18, which digit is in the tenths place?

A. 1

B. 2

C. 4

D. 8

8. Which decimal has the digit 4 in the hundreds place and in the tenths place?

A. 342.49

B. 418.47

C. 436.34

D. 574.48

9. The length of Scott's baseball bat is 0.81 meters. Write this decimal in word form.

Answer _____

10. Larry spent five dollars and forty-eight cents on a new baseball. Write this amount in standard form.

Answer _____

EXTENDED-RESPONSE QUESTION

11. Mr. Tyler drew the following grids on the board.

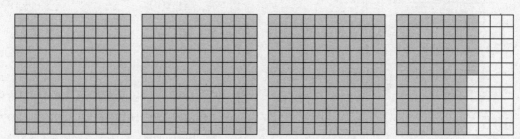

Part A Write the decimal that Mr. Tyler's model represents in expanded form.

_____

Part B Mr. Tyler said that his model represents the cost of a new pen. Write the money amount that his model represents.

_____

17 Compare and Order Decimals

4.N.12, 4.A.2

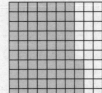

Getting the Idea

Comparing decimals is similar to comparing whole numbers. Use the same symbols to compare.

The symbol > means **is greater than**.

The symbol < means **is less than**.

The symbol = means **is equal to**.

The symbol   means **is not equal to**.

EXAMPLE 1

Which symbol makes this statement true?

$$\$0.76 \ \bigcirc \ \$0.74$$

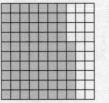

STRATEGY   Line up the numbers on the decimal point. Start comparing the digits in the greatest place.

STEP 1   Line up the numbers on the decimal point.

0.76

0.74

STEP 2   Compare the digits in the ones place.

0.76

0.74

Since 0 ones = 0 ones, compare the next greatest place.

STEP 3   Compare the digits in the tenths place.

0.76

0.74

Since 7 tenths = 7 tenths, compare the next greatest place.

STEP 4   Compare the digits in the hundredths place.

0.76

0.74

6 hundredths > 4 hundredths

0.76 > 0.74

SOLUTION $0.76 > $0.74

EXAMPLE 2

Lillian's decimal grid had 0.32 shaded. Emily's decimal grid had 0.34 shaded. Which number is greater?

STRATEGY   **Use a number line to compare the decimals.**

STEP 1   Draw a number line from 0.30 to 0.40.

Put dots at 0.32 and 0.34.

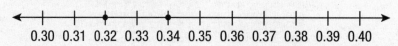

0.30 0.31 0.32 0.33 0.34 0.35 0.36 0.37 0.38 0.39 0.40

STEP 2   Compare the placement of the two decimals on the number line.

0.32 is farther to the left than 0.34.

So 0.32 is less than 0.34.

STEP 3   Use < or > to compare the decimals.

0.32 < 0.34

SOLUTION   **0.32 < 0.34**

EXAMPLE 3

Order the following money amounts from greatest to least.

$0.63 $0.70 $0.67

STRATEGY   **Use a place-value chart to compare the numbers.**

STEP 1   Write the money amounts in a place-value chart.

Ones		Tenths	Hundredths
0	.	6	3
0	.	6	7
0	.	7	0

STEP 2   Compare the digits in the tenths place.

7 tenths > 6 tenths

So $0.70 is the greatest money amount.

STEP 3   Compare the digits in the hundredths place.

7 hundredths > 3 hundredths

So $0.67 is greater than $0.63.

SOLUTION   **The money amounts ordered from greatest to least are $0.70, $0.67, and $0.63.**

Putting a 0 at the end of a decimal does not change its value.

For example, 0.5 = 0.50 and 2.8 = 2.80.

COACHED EXAMPLE

Order the following decimals from greatest to least.

$0.4 $0.39 $0.52 $0.48

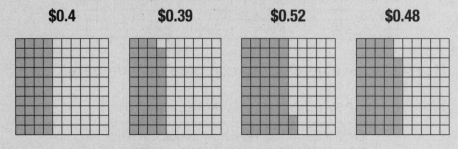

THINKING IT THROUGH

Use place value to compare the numbers.

Write the decimals in a place-value chart.

Ones	.	Tenths	Hundredths

Compare the digits in the greatest place: _____.

All of the digits are the _____.

Compare the digits in the next greatest place: _____.

_____ tenths > _____ tenths > _____ tenths

So the greatest decimal is _____ and the least decimal _____.

To compare the remaining decimals, compare the digits in the next greatest place: _____.

_____ hundredths > _____ hundredths

So _____ > _____

The decimals in order from greatest to least are _____;
_____; _____; _____.

Lesson Practice

Choose the correct answer.

1. Which symbol belongs in the circle?

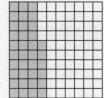

$$\$0.38 \bigcirc \$0.36$$

   A. >   **C.** =

   B. <   **D.**

2. Which money amount makes this number sentence true?

$$\$0.28 < \square$$

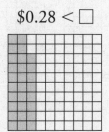

   A. $0.03   **C.** $0.27

   B. $0.18   **D.** $0.32

3. Which correctly compares $0.85 and $0.87?

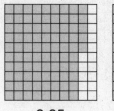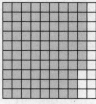

       0.85   0.87

   A. $0.85 = $0.87

   B. $0.85 < $0.87

   C. $0.85 > $0.87

   D. $0.85 + $0.87

4. Which lists the decimals in order from greatest to least?

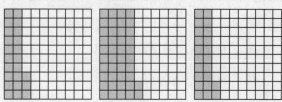

   A. 0.23, 0.42, 0.22

   B. 0.22, 0.23, 0.42

   C. 0.42, 0.23, 0.22

   D. 0.22, 0.42, 0.23

5. Which correctly compares $0.39 and $0.34?

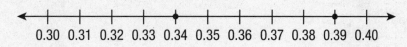

   0.30 0.31 0.32 0.33 0.34 0.35 0.36 0.37 0.38 0.39 0.40

   A. $0.39 > $0.34

   B. $0.39 < $0.34

   C. $0.39 = $0.34

   D. $0.39 × $0.34

6. Which lists the money amounts in order from greatest to least?

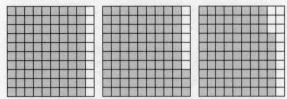

A. $0.87, $0.93, $0.90

B. $0.90, $0.87, $0.93

C. $0.93, $0.90, $0.87

D. $0.87, $0.90, $0.93

7. Which lists the money amounts in order from least to greatest?

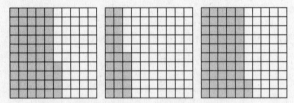

A. $0.54, $0.25, $0.52

B. $0.54, $0.52, $0.25

C. $0.52, $0.25, $0.54

D. $0.25, $0.52, $0.54

8. Which symbol makes this sentence true?

$0.78 ◯ $0.78

A. <   C. =

B. >   D.

9. George bought a piece of gum for $0.50 and Tommy bought a piece of licorice for $0.49. Who paid more for their piece of candy?

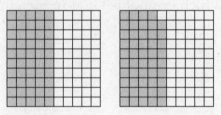

Answer _____

10. After buying cookies at the bake sale, Spencer had $0.25 left over and Talia had $0.35 left over. Who had less money left over?

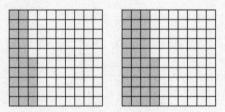

Answer _____

18 Add and Subtract Decimals

4.N.25

Getting the Idea

You can use models to add and subtract decimals.

EXAMPLE 1

Find the sum: 0.57 + 0.33

STRATEGY   **Make a model to add.**

    STEP 1   Use a grid to model 0.57.

    STEP 2   Use the same model to add 0.33.

    STEP 3   Write the total number of shaded squares as a decimal.

        0.90

SOLUTION   0.57 + 0.33 = 0.90

EXAMPLE 2

Mark bought 1.15 pounds of bananas and 0.67 pound of grapes. How many pounds of fruit did Mark buy in all?

STRATEGY   **Use models to add 1.15 + 0.67.**

STEP 1   Use grids to model 1.15.

Use two grids and shade the first one completely.

1.15 is *one and fifteen hundredths*, so shade 15 squares in the second grid.

 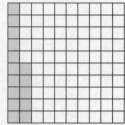

STEP 2   Use the same model to add 0.67.

0.67 is *sixty-seven hundredths*, so shade 67 more squares in the second grid.

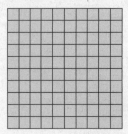 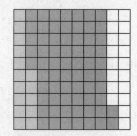

STEP 3   Write the total number of shaded parts as a decimal.

1.82

SOLUTION   **Mark bought 1.82 pounds of fruit in all.**

EXAMPLE 3

Susan recorded 1.6 inches of rainfall from the last storm. Her friend Anna, on the other side of town, recorded 0.85 inches from the same storm. How much more rainfall did Susan record than Anna?

STRATEGY   Use models to subtract 1.6 − 0.85.

    STEP 1   Model the greater decimal using grids.

        Use two grids and shade the first one completely.

        1.6 is *one and six tenths*, or *one and sixty hundredths*. Shade 60 squares in the second grid.

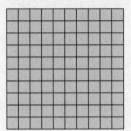 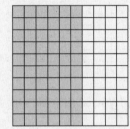

    STEP 2   Cross out squares to represent the number being subtracted.

        0.85 is *eighty-five hundredths*, so cross out 85 of the shaded squares.

        Cross out 60 shaded squares in the second grid. Cross out 25 more in the first grid.

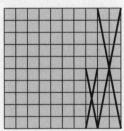 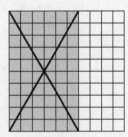

    STEP 3   Count the number of shaded squares that are not crossed out.

        75 squares are shaded and not crossed out.

        75 hundredths = 0.75

SOLUTION   **Susan recorded 0.75 inch more rainfall than Anna.**

COACHED EXAMPLE

Derek walked 0.75 mile from his house to the library. Then he walked 0.6 mile to Fred's house. How many miles did Derek walk in all?

THINKING IT THROUGH

Do you need to add or subtract? _____

Shade squares in the model to show how far Derek walked from his house to the library.

How many squares did you shade to show the distance from Derek's house to the library? _____

How many more squares need to be shaded to show the distance from the library to Fred's house? _____

How many squares can be shaded in the grid on the left? _____

How many squares do you need to shade in the grid on the right? _____

What decimal represents the total number of shaded parts? _____

Derek walked _____ miles in all.

Lesson Practice

Choose the correct answer.

1. What is 1.25 + 0.65?

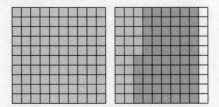

- **A.** 1.7
- **B.** 1.8
- **C.** 1.9
- **D.** 1.95

2. What is 0.93 − 0.47?

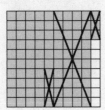

- **A.** 0.44
- **B.** 0.46
- **C.** 0.47
- **D.** 0.56

3. What is 2.7 + 0.24?

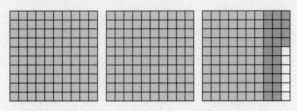

- **A.** 2.31
- **C.** 2.84
- **B.** 2.46
- **D.** 2.94

4. What is 1.4 − 0.89?

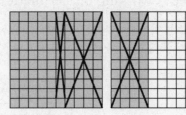

- **A.** 0.23
- **B.** 0.41
- **C.** 0.51
- **D.** 0.61

5. What is 0.36 + 0.19?

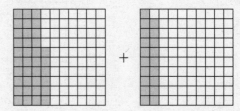

- **A.** 0.45
- **B.** 0.46
- **C.** 0.55
- **D.** 0.56

6. What is 0.87 − 0.39?

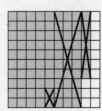

- **A.** 0.38
- **B.** 0.42
- **C.** 0.48
- **D.** 0.52

EXTENDED-RESPONSE QUESTION

7. Shaylee is working on her math homework. Help her add and subtract the decimals shown below.

   Part A   Shade the grids below to show 1.24 + 0.43.

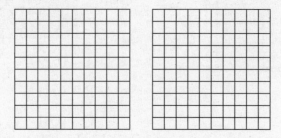

   Part B   Shade and use cross-out markings on the grids below to show 1.82 − 0.74.

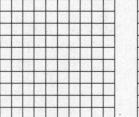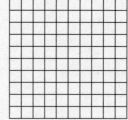

Lesson

19 Fraction-Decimal Equivalencies

 4.N.24

Getting the Idea

Fractions and decimals are related. For example, the shaded part of the figure below can be expressed as both a decimal and a fraction.

$$0.7 = \textit{seven tenths} = \frac{7}{10}$$

To change a decimal to a fraction, write the decimal as the numerator over the denominator of the least decimal place. For example, $0.25 = \frac{25}{100}$.

EXAMPLE 1

What fraction is equal to 0.6?

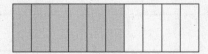

STRATEGY   **Write the decimal as a fraction.**

  STEP 1   Find the numerator and the denominator.

              The numerator is 6. The decimal is in tenths, so the denominator is 10.

  STEP 2   Write the fraction.

              $\frac{\text{numerator}}{\text{denominator}} \rightarrow \frac{6}{10}$

SOLUTION $0.6 = \frac{6}{10}$

EXAMPLE 2

What fraction is equal to 0.04?

STRATEGY   **Write the decimal as a fraction.**

STEP 1   Find the numerator and the denominator.

The numerator is 4. The decimal is in hundredths, so the denominator is 100.

STEP 2   Write the fraction.

$$\frac{numerator}{denominator} \rightarrow \frac{4}{100}$$

SOLUTION $0.04 = \frac{4}{100}$

EXAMPLE 3

What fraction is equal to 0.2?

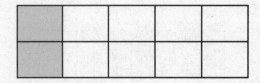

STRATEGY   **Write the decimal as a fraction.**

STEP 1   Find the numerator and the denominator.

The numerator is 2. The decimal is in tenths, so the denominator is 10.

STEP 2   Write the fraction.

$$\frac{numerator}{denominator} \rightarrow \frac{2}{10}$$

SOLUTION $0.2 = \frac{2}{10}$

EXAMPLE 4

What decimal names the shaded part of the grid?

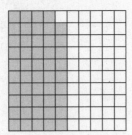

STRATEGY   Use what you know about fractions to find the decimal.

STEP 1   Write the fraction that represents the shaded part of the grid.

49 squares are shaded. There are 100 squares in all.

The fraction is $\frac{49}{100}$.

STEP 2   Write the decimal that is equal to the fraction $\frac{49}{100}$.

0.49 is equal to $\frac{49}{100}$.

SOLUTION   The decimal 0.49 names the shaded part of the grid.

COACHED EXAMPLE

What fraction and decimal represent the part of the grid that is shaded?

THINKING IT THROUGH

There are _____ squares in the grid. _____ squares are shaded.

The fraction _____ represents the shaded part of the grid.

The fraction _____ is equal to the decimal _____.

Both _____ and _____ represent the part of the grid that is shaded.

Lesson Practice

Choose the correct answer.

1. Which is 0.60 written as a fraction?

   A. $\frac{4}{10}$

   B. $\frac{4}{100}$

   C. $\frac{60}{10}$

   D. $\frac{60}{100}$

2. Which is 0.8 written as a fraction?

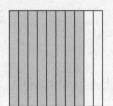

   A. $\frac{2}{8}$

   B. $\frac{2}{10}$

   C. $\frac{8}{10}$

   D. $\frac{8}{100}$

3. Rod's puppy ate $\frac{80}{100}$ of a can of dog food. Which model shows the amount the puppy ate?

   A.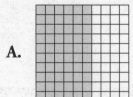

   B.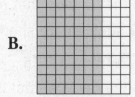

   C.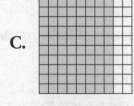

   D.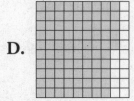

4. Which does **not** equal 0.7?

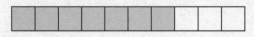

   A. $\frac{7}{10}$

   B. $\frac{10}{7}$

   C. 0.70

   D. $\frac{70}{100}$

5. Which of the following has the same value as 0.19?

   A. $\frac{10}{19}$

   B. $\frac{19}{100}$

   C. $\frac{19}{10}$

   D. $\frac{100}{19}$

6. What fraction is equal to 0.3?

   A. $\frac{3}{100}$

   B. $\frac{3}{10}$

   C. $\frac{1}{3}$

   D. $\frac{10}{3}$

7. Wendy paid $0.08 for a pencil. What fraction of a dollar represents $0.08?

   Answer _____

8. Hyun helps her dad cut up firewood. She cuts up 0.4 of a tree. What fraction represents the amount of the tree Hyun cut?

   Answer _____

EXTENDED-RESPONSE QUESTION

9. Darius has $0.54 in his pocket.

   Part A Shade the model to show $0.54.

   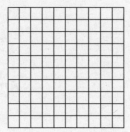

   Part B What fraction has the same value as $0.54?

   _____

20 Solve Real-World Problems

4.N.15

Getting the Idea

To solve real-world problems, look for key words that tell you whether to add, subtract, multiply, or divide.

Add to find the total when given two or more values. Key words to look for are *more*, *altogether*, and *in all*.

Subtract to find the value that remains when an amount is removed. Key words to look for are *fewer*, *take away*, and *is left over*.

Multiply to find the total when given equal groups. Key words to look for are *each*, *per*, and *in all*.

Divide to find the number of equal groups, or the number in equal groups. Key words to look for are *groups of* and *equal number*.

You can use a variety of different computational methods, such as paper and pencil, mental math, or a calculator, to solve real-world problems.

EXAMPLE 1

Barbara has 4 more video games than John. John has 3 fewer video games than Larry. Larry has 13 video games. How many video games does Barbara have?

STRATEGY   Decide which operation to use. Then solve the problem.

STEP 1   Decide which operations to use.

The key words in the problem are *more* and *fewer*.

I need to use both addition and subtraction.

STEP 2   First find how many video games John has using subtraction.

Larry's video games − 3 = John's video games

13 − 3 = 10

STEP 3   Then find how many video games Barbara has using addition.

John's video games + 4 = Barbara's video games

10 + 4 = 14

SOLUTION   **Barbara has 14 video games.**

EXAMPLE 2

Vera makes $15 per hour when she babysits. How much will she make in all if she works for 8 hours?

STRATEGY   **Decide which operation to use. Then solve the problem.**

STEP 1   Decide which operation to use.

The key words in the problem are *per hour* and *in all*.

The problem asks for a total, so multiply.

STEP 2   Multiply.

$15 \times 8 = 120$

SOLUTION   **Vera will make $120.**

EXAMPLE 3

There are 54 students going on a field trip to the zoo. They are taking 3 buses to the zoo. Each bus holds an equal number of students. How many students will ride on each bus?

STRATEGY   **Decide which operation to use. Then solve the problem.**

STEP 1   Decide which operation to use.

The key words in the problem are *each* and *equal number*.

The problem asks for the number of students on each bus, so divide.

STEP 2   Divide.

$54 \div 3 = 18$

SOLUTION   **18 students will ride on each bus.**

COACHED EXAMPLE

A worker filled boxes with 96 bottles of water. He filled 4 boxes with an equal number of bottles. How many bottles of water are in each box?

THINKING IT THROUGH

What are the key words? _____

What operation should you use? _____

Solve the problem below.

There are _____ bottles of water in each box.

Lesson Practice

Choose the correct answer.

1. Hiro brushes his teeth 3 times each day. How many times will he brush his teeth in all in 31 days?

   A. 10

   B. 11

   C. 93

   D. 930

2. Lilly is 7 years younger than Ricky. Ricky is 9 years older than Thomas. Thomas is 17 years old. How old is Lilly?

   A. 26

   B. 19

   C. 15

   D. 10

3. Mr. Parker has 24 students. Each student has 9 pages of social studies reading to do before the end of the day. How many pages of social studies reading do the students have to do in all?

   A. 186

   B. 206

   C. 216

   D. 316

4. There are 85 pencils in the classroom supply drawer. Each student is given 5 pencils from the drawer. How many students each receive 5 pencils?

   A. 17

   B. 18

   C. 19

   D. 27

5. There are 96 students in the fourth grade at Misty Hill Elementary School. Each of the students took 5 tests this week. How many tests did the students take in all?

   A. 330

   B. 380

   C. 430

   D. 480

6. Peter travels a total of 105 miles to and from work each week. He works 5 days each week. How many miles does Peter travel each day?

   A. 525

   B. 210

   C. 21

   D. 15

7. Julia and Catherine both collect stickers. Julia has collected 58 stickers and Catherine has collected 66. How many stickers have they collected in all?

   A. 134
   B. 124
   C. 114
   D. 104

8. The local clothing shop sells T-shirts for $15 each. It sold 8 T-shirts on Thursday. Which is the best operation to use to find how much money the store made selling T-shirts on Thursday?

   A. addition
   B. subtraction
   C. multiplication
   D. division

9. Samantha received a 95 on the math test. Franz received an 87 on the math test. How many more points did Samantha receive on the math test than Franz?

   Answer _____

10. Tracy and her 2 sisters went apple picking. They each picked 15 apples. Write a number sentence to show how many apples Tracy and her sisters picked in all.

    Answer _____

21 Use Estimation to Check Answers

4.N.27

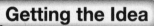 **Getting the Idea**

An **estimate** is a number that is close to the exact amount. Estimation is a way to solve problems when an exact answer is not needed. The word *about* is a clue that you should estimate the answer.

To estimate solutions to a problem, round the numbers given in the problem. Then use the rounded numbers to compute an estimated answer. For a review of how to round numbers, look at Lesson 3.

EXAMPLE 1

Toby works at a pizza parlor. He earned $73 the first week, $102 the second week, and $86 the third week. About how much money did Toby earn in 3 weeks?

STRATEGY Round each amount to the nearest ten and then add.

STEP 1 Round each money amount to the nearest 10.

$73 rounds down to $70.

$102 rounds down to $100.

$86 rounds up to $90.

STEP 2 Add the rounded numbers.

70 + 100 + 90 = 260

SOLUTION Toby earned about $260 in 3 weeks.

Estimation is also a way to decide if an answer is reasonable. By estimating, you can quickly tell if you made a place-value error when computing.

EXAMPLE 2

Jane is reading a 232-page book. She has read 113 pages. Jane says she has 219 pages left to read. Is her answer reasonable?

STRATEGY   **Estimate the answer. Then compute.**

STEP 1   Round each number of pages to the nearest 100.

232 rounds down to 200.

113 rounds down to 100.

STEP 2   Subtract the rounded numbers.

$200 - 100 = 100$

STEP 3   Compare the estimate to Jane's answer.

100 is not close to 219.

SOLUTION   **Jane's answer of 219 pages is not reasonable.**

To estimate the product of a 2-digit number and a 1-digit number, round the 2-digit number to the nearest ten, then multiply it by the 1-digit number.

EXAMPLE 3

Phil delivers 48 newspapers each day of the week. Phil says that he delivered 336 newspapers last week. Is Phil's answer reasonable?

STRATEGY   **Round the 2-digit number to the nearest ten and multiply.**

STEP 1   Round the 2-digit number to the nearest ten.

48 rounds up to 50.

STEP 2   Multiply by 7 since there are 7 days in a week.

$50 \times 7 = 350$

STEP 3   Compare the estimate to Phil's answer.

350 is close to 336.

SOLUTION   **Phil's answer of 336 is reasonable.**

When estimating products and quotients, use **compatible numbers**. Compatible numbers are numbers close to the actual numbers you need to multiply or divide that are easy to compute with mentally.

EXAMPLE 4

Adele orders 95 roses for a party. If she has 9 vases, about how many roses will Adele put in each vase?

STRATEGY   Choose compatible numbers close to the actual numbers. Then divide.

  STEP 1   Choose compatible numbers.

  95 is close to 100.

  9 is close to 10.

  STEP 2   Divide using the compatible numbers.

  $100 \div 10 = 10$

SOLUTION   **Adele will put about 10 roses in each vase.**

COACHED EXAMPLE

There are 52 weeks in a year. Today is Cindy's ninth birthday. Cindy says she has lived 638 weeks in 9 years. Is Cindy's answer reasonable?

THINKING IT THROUGH

Choose compatible numbers.

_____ is close to 52.

_____ is close to 9.

_____ using the compatible numbers.

_____ × _____ = _____

Is _____ close to 638?_____

Cindy's answer _____ reasonable.

Lesson Practice

Choose the correct answer.

1. Leon has 291 baseball cards in his collection. Cora has 533 baseball cards in her collection. About how many more baseball cards does Cora have than Leon?

    A. 400

    B. 300

    C. 200

    D. 100

2. Amy wants to buy a skirt that costs $28, a blouse that costs $24, and a purse that costs $17. Which is the best estimate for how much Amy will spend?

    A. $70

    B. $60

    C. $50

    D. $40

3. Reggie and Tony collected canned goods for a food drive. Reggie collected 32 cans and Tony collected 46 cans. Which expression could be used to estimate how many cans Reggie and Tony collected in all?

    A. 40 + 40

    B. 40 + 50

    C. 30 + 40

    D. 30 + 50

4. There were 78 visitors to a museum exhibit on one day. If the same number of people visited each day, what is the best estimate for the total number of visitors after 8 days?

    A. 540

    B. 640

    C. 740

    D. 840

5. For which of the following expressions is 100 the best estimate?

    A. 39×4

    B. 46×2

    C. 57×4

    D. 67×2

6. The local bakery makes 7 trays of oatmeal cookies each morning. Each tray holds 22 cookies. About how many cookies does the bakery make each morning?

    A. 140

    B. 180

    C. 210

    D. 300

7. Last summer, Terrance earned $1,403 and Aaliyah earned $875. About how much more money did Terrance earn than Aaliyah?

   A. $600

   B. $500

   C. $400

   D. $300

8. Which is the best estimate for 45 ÷ 7?

   A. 4

   B. 6

   C. 8

   D. 9

9. The fourth-grade class performed 3 shows for a total of 91 people. Each show had about the same number of people. About how many people attended each show?

   Answer_____

10. There are 6 bins of aluminum cans to be recycled. Each bin has 82 cans. About how many cans are there in all?

    Answer_____

1 Review

1 What is the next number in this pattern?

3,625   4,625   5,625   _____

A 5,626

B 5,635

C 5,725

D 6,625

2 Which statement is true?

A 4,154 > 4,514

B 5,392 > 5,329

C 6,081 < 6,018

D 7,260 = 7,620

3 Each fourth-grade teacher at Samara's school has 26 students. There are 12 fourth-grade teachers. Samara estimates that there are about 300 fourth-grade students at her school. Which expression can be used to check whether Samara's estimate is reasonable?

A 20×10

B 20×20

C 30×10

D 30×20

4 George wants to check that he correctly solved the number sentence below.

$18 \div 3 = 6$

Which number sentence could George use to check to see if his answer is correct?

A $6 \times 3 = \square$

B $6 \div 3 = \square$

C $6 \times 18 = \square$

D $18 \times 3 = \square$

5 How is 6,038 written in expanded form?

A $600 + 30 + 8$

B $6,000 + 30 + 8$

C $6,000 + 300 + 8$

D $6,000 + 300 + 80$

6 The circus had an attendance of four thousand, sixty-eight people. How is that number written in standard form?

A 4,068

B 4,608

C 4,680

D 8,604

7 Caroline is reading a book that is 462 pages long. What is 462 rounded to the nearest ten?

A 500

B 470

C 460

D 400

8 Which number when multiplied by any odd number **always** results in an odd number?

A 2

B 4

C 6

D 7

9 Which number makes this number sentence true?

$$4 \times (5 \times 3) = (4 \times 5) \times \underline{\quad}$$

A 3

B 4

C 5

D 60

10 Which expression means the same as 6×4?

A $6 + 6 + 6 + 6$

B $4 + 6$

C $6 + 6 + 6 + 6 + 6$

D $4 + 4 + 4 + 4 + 4 + 4 + 4$

11 What is the sum of $0.27 + 0.46$?

A 0.63

B 0.73

C 0.83

D 0.87

12 82 ÷ 3 =

A   20 R2

B   24

C   26 R2

D   27 R1

13 Which is the same as
6 thousands + 5 tens +
four ones?

A   654

B   6,054

C   6,540

D   65,004

14 How is eight dollars and
forty-two cents written as
a decimal?

A $8.42

B $8.24

C $2.84

D $2.48

15 Karim walked $\frac{3}{10}$ mile to Fred's
apartment. Then they walked
$\frac{4}{10}$ mile to the park. How far did
Karim walk all together?

A $\frac{1}{10}$ mile

B $\frac{7}{20}$ mile

C $\frac{7}{10}$ mile

D $\frac{8}{10}$ mile

16 Which rectangle shows **exactly**
$\frac{1}{3}$ shaded?

A

B

C

D

17 What is the product of 32 × 24?

A   192

B   720

C   768

D   868

18 Which fraction represents point D on the number line?

    A $\frac{3}{8}$

    B $\frac{1}{2}$

    C $\frac{5}{8}$

    D $\frac{3}{4}$

19 It rained on 0.7 of the days of Omar's vacation. What fraction of the days did it rain?

    A $\frac{3}{100}$

    B $\frac{7}{100}$

    C $\frac{3}{10}$

    D $\frac{7}{10}$

20 Which orders the money amounts from **least** to **greatest**?

    A $0.25, $0.31, $0.36

    B $0.31, $0.36, $0.25

    C $0.36, $0.25, $0.31

    D $0.25, $0.36, $0.31

21 Which sentence is true?

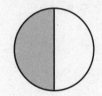

    A $\frac{1}{2} > \frac{1}{4}$

    B $\frac{1}{2} = \frac{1}{4}$

    C $\frac{1}{2} < \frac{1}{4}$

    D $\frac{1}{4} > \frac{1}{2}$

22 Each CD that Rashaun bought last year cost $16. He bought 42 CDs last year. How much money did Rashaun spend on CDs last year?

Show your work.

Answer _____

23 Ynez has 43 DVDs in her collection. She has 8 DVDs on each shelf except for 1 shelf. How many DVDs are on the other shelf?

Show your work.

Answer _____ DVDs

24 Marcus plans to read 32 pages each day. It will take him 7 days to finish the book if he follows his plan. How many pages long is Marcus's book?

Show your work.

Answer _____ pages

25 The table shows the number of points that four players earned in a game.

PLAYER SCORES

Player	Number of Points
Jill	3,423
Manny	2,875
Eric	4,148
Karen	3,039

Part A

How many points have Jill and Karen scored altogether?

Show your work.

Answer _____ points

Part B

How many more points did Eric score than Manny?

Show your work.

Answer _____ points

26   The table shows the mean radius in miles of four planets.

MEAN RADIUS OF PLANETS

Planet	Mean Radius (in miles)
Earth	3,958
Mars	2,106
Mercury	1,516
Venus	3,760

Part A

Order the planets using the mean radii in miles from **greatest** to **least**.

Answer _____

Part B

On the lines below, explain how you found your answer.

_____

_____

_____

_____

STRAND

2 Algebra

			NYS Math Indicators
Lesson 22	Open Sentences with Equations	140	4.A.1
Lesson 23	Open Sentences with Inequalities	143	4.A.1, 4.A.2, 4.A.3
Lesson 24	Number and Geometric Patterns	148	4.A.4
Lesson 25	Input-Output Tables	155	4.A.5
Strand 2 Review		161	

22 Open Sentences with Equations

4.A.1

Getting the Idea

An **open sentence** is a number sentence that has a symbol to represent a missing value. An open sentence can be an **equation**, which are number sentences that contain an equals sign. Here are some examples of open sentences that are equations.

$$9 + \square = 17 \qquad 54 \times 2 = \triangle \qquad \underline{\qquad} = 7 \times 3$$

A **variable** is a letter or symbol that stands for a number that can change or vary. To solve an open sentence, find the value or values that makes the sentence true.

You can use an open sentence to represent a real-world situation. To do this, you need to choose the correct operation (addition, subtraction, multiplication, or division) that would be needed to solve the problem.

EXAMPLE 1

Whitney needs to buy 36 buttons for a craft project. The buttons are sold in packages of 4. Write an open sentence that models this situation.

STRATEGY Find the key words.

STEP 1 Look for a clue about which operation to use.

The phrase "packages of 4" means you are looking for how many groups of 4. You need to divide.

STEP 2 Find the number facts in the problem.

Whitney needs 36 buttons.

The buttons are sold in packages of 4.

STEP 3 Write an open sentence using $\square$ for the missing value.

$36 \div 4 = \square$

SOLUTION The open sentence that models this situation is $36 \div 4 = \square$.

To solve an open sentence, find the value or values that make the sentence true. Sometimes you can use mental math to solve for the unknown value.

EXAMPLE 2

What value makes this open sentence true?

$\square + 7 = 15$

STRATEGY   Use mental math.

Think: What number plus 7 is 15?

$8 + 7 = 15$

SOLUTION   The value that makes the open sentence $\square + 7 = 15$ true is 8.

COACHED EXAMPLE

Pauline jogged a total of 9 miles on Friday and Saturday. If she jogged 4 miles on Friday, how far did she jog on Saturday? Write an open sentence that models the situation.

THINKING IT THROUGH

The key words "a total of" tell you to use _____.

Find the number facts in the problem.

Pauline jogged _____ miles on Friday.

She jogged a total of _____ miles on Friday and Saturday.

The missing value is how many miles Pauline jogged on _____.

Write an open sentence using $\triangle$ to represent the missing value. Use $\triangle$ for the missing value.

_____

The open sentence that models this situation is _____.

Lesson Practice

Choose the correct answer.

1. There are 25 students in Mr. Palmer's class. Of his students, 14 are boys. Which open sentence could be used to find how many students are girls?

   A. $25 + 14 = \square$

   B. $14 \times \square = 25$

   C. $14 + \square = 25$

   D. $\square - 25 = 14$

2. What value makes this open sentence true?

   $$6 \times \square = 36$$

   A. 6

   B. 30

   C. 42

   D. 216

3. What value makes this open sentence true?

   $$12 + \square = 30$$

   A. 8

   B. 18

   C. 28

   D. 42

4. The county basketball league divided its 90 players into equal teams with 5 players on each team. Which open sentence could be used to find the number of teams?

   A. $90 = 5 + \square$

   B. $90 \div 5 = \square$

   C. $90 - \square = 5$

   D. $\square \div 5 = 90$

5. Antonio has a total of 24 rap and rock CDs. Eight of the CDs are rap music. Write an open sentence that could be used to find the number of rock CDs in Antonio's collection.

   Answer _____

6. Mrs. Colton wants to decorate her bulletin board with flowers. Each flower will need 5 petals. She asks her class to cut out 60 petals. Write an open sentence that could be used to show how many flowers Mrs. Colton wants for the bulletin board.

   Answer _____

23 Open Sentences with Inequalities

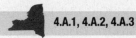
4.A.1, 4.A.2, 4.A.3

Getting the Idea

An open sentence can also be an **inequality** which is a number sentence that states that two sides of an equation are not equal.

Here are some examples of open sentences that are inequalities.

$$19 - \square < 7 \qquad \_\_\_\_\_ > 2 \times 6$$

When you are writing an inequality, you need to choose a comparing symbol: $<$ or $>$.

EXAMPLE 1

It takes Joel less than 18 minutes to walk to school from his house. He left home 4 minutes ago. Write an open sentence that tells how many more minutes it should take Joel to get to school.

STRATEGY **Write an open sentence that models the situation.**

STEP 1   Look for a clue about which operation to use.

   The key words "how many *more* minutes" tell you to use addition.

STEP 2   Determine the comparing symbol to use.

   Since it takes Joel *less* than 18 minutes, you need to use the $<$ symbol.

STEP 3   Identify the number facts given in the problem.

   Joel has been walking for 4 minutes.

   $\square$ is the amount of minutes Joel has left to walk.

   His total walking time will be less than 18 minutes.

SOLUTION $4 + \square < 18$

To solve an open sentence, find the value or values that makes the sentence true.

Like an equation, an inequality can be made true or false by replacing the missing value with a number. For the inequality □ + 5 < 9, if □ is replaced with 3 or any whole number less than 3, then the inequality is a true sentence. Any number greater than 3 makes the inequality false.

EXAMPLE 2

What values for x will make this open sentence true?

12 × □ > 48

STRATEGY   **Solve for □. The solution will make the sentence true.**

STEP 1   What number times 12 is equal to 48?

4 × 12 = 48

Any number greater than 4 will make the sentence true.

STEP 2   Check your answer.

12 × 5 = 60

60 > 48

Your answer is correct.

SOLUTION   **When the values for □ > 4, the open sentence is true.**

Sometimes you can determine if an open sentence is true or false without substituting a numeric value for each variable.

EXAMPLE 3

Mr. Philips wrote the following open sentences on the board.

If both of Mr. Philips's open sentences are true, which open sentence is also true?

A. ◯ > ☾

B. ☾ < ◇

C. ◇ < ☾

D. ☾ < ◯

STRATEGY   **Check each answer choice to find the true open sentence.**

STEP 1   Look at choice A. ◯ > ☾

Choice A is not true because ◯ < ☾.

STEP 2   Look at choice B. ☾ < ◇

Choice B is not true because ◇ < ☾ and ◯ < ☾.

STEP 3   Look at choice C. ◇ < ☾

Choice C is true because ◇ < ◯ and ◯ < ☾.

STEP 4   Look at choice D. ☾ < ◯

Choice D is not true because ◯ < ☾.

SOLUTION   **Choice C is a true open sentence.**

COACHED EXAMPLE

What values for ☐ will make this open sentence true?

$$25 - \square < 5$$

THINKING IT THROUGH

Find the number that makes the sentence below true.

$25 -$ _____ $= 5$

So any number greater than _____ will make the open sentence true.

Check your answer. Choose a number greater than 20.

$25 -$ _____ < 5

_____ < 5

Is this sentence true? _____

Any value greater than _____ will make the open sentence true.

Lesson Practice

Choose the correct answer.

1. What value for □ will make this open sentence false?

    $7 + □ > 15$

    A. 12

    B. 10

    C. 9

    D. 7

2. What value for □ will make this open sentence true?

    $5 × □ < 25$

    A. 4

    B. 5

    C. 6

    D. 7

3. Mrs. Peroni's chickens laid 14 eggs in one week. By the end of the second week, Mrs. Peroni's chickens had laid more than 25 eggs. Which open sentence could be used to show how many eggs the chickens laid in the second week?

    A. $14 − □ > 25$

    B. $14 + □ > 25$

    C. $14 × □ < 25$

    D. $14 ÷ □ > 25$

4. Miss McClinsey bought more than 15 yards of fabric to make banners for the parade. When she finished making the banners, she had 2 yards of fabric left over. Which open sentence could be used to find how much fabric she used for the banners?

    A. $□ × 2 < 15$

    B. $15 ÷ 2 = □$

    C. $□ + 2 > 15$

    D. $2 + □ = 15$

5. The following open sentences are true.

    $□ < ○$ and $○ < \#$

    Which open sentence is also true?

    A. $\# < ○$

    B. $□ > ○$

    C. $\# > □$

    D. $○ > \#$

6. What value for □ will make this open sentence true?

    $6 × □ < 30$

    Answer_____

7. $△ > □$ and $□ > ○$ are true open sentences. Write another open sentence that is also true.

    Answer_____

24 Number and Geometric Patterns

4.A.4

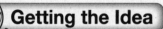
Getting the Idea

A **pattern** is an ordered group of numbers or shapes arranged according to a **rule**. The rule for a pattern tells how the numbers or shapes are related. The numbers in a pattern can increase through addition or multiplication. The numbers can decrease through subtraction or division.

EXAMPLE 1

What is a rule for this pattern? 6, 18, 54, 162, …

STRATEGY   Find how the numbers are changing.

STEP 1   Decide if the numbers increase or decrease.

The numbers increase.

STEP 2   Since the numbers are increasing, use addition or multiplication.

Try addition.

$6 + ? = 18 \quad \rightarrow \quad 6 + 12 = 18$

$18 + ? = 54 \quad \rightarrow \quad 18 + 36 = 54$

$54 + ? = 162 \quad \rightarrow \quad 54 + 108 = 162$

The numbers increase by different amounts, so it is not addition.

STEP 3   Try multiplication.

$6 \times ? = 18 \quad \rightarrow \quad 6 \times \mathbf{3} = 18$

$18 \times ? = 54 \quad \rightarrow \quad 18 \times \mathbf{3} = 54$

$54 \times ? = 162 \quad \rightarrow \quad 54 \times \mathbf{3} = 162$

Each number is 3 times the previous number.

STEP 4   Identify a rule for the pattern.

A rule is multiply by 3.

SOLUTION   A rule for the pattern is multiply by 3.

EXAMPLE 2

What is the eighth number in this pattern?

15, 22, 29, 36, …

STRATEGY   **Find a rule for the pattern. Then extend the pattern.**

STEP 1   Decide if the numbers increase or decrease.

The numbers increase.

STEP 2   Find a rule for the pattern.

Since the numbers are increasing, use addition or multiplication.

Try addition.

$15 + ? = 22 \quad \rightarrow \quad 15 + \mathbf{7} = 22$

$22 + ? = 29 \quad \rightarrow \quad 22 + \mathbf{7} = 29$

$29 + ? = 36 \quad \rightarrow \quad 29 + \mathbf{7} = 36$

A rule is add 7.

STEP 3   Use the rule to extend the pattern.

fifth number: $36 + 7 = 43$

sixth number: $43 + 7 = 50$

seventh number: $50 + 7 = 57$

eighth number: $57 + 7 = 64$

SOLUTION   **The eighth number in the pattern is 64.**

EXAMPLE 3

The table shows how much money Mateo had in his school lunch account from Monday through Thursday.

Monday	Tuesday	Wednesday	Thursday	Friday
$150	$135	$120	$105	?

How much money will Mateo have in his account on Friday?

STRATEGY   **Find a rule for the pattern. Then find the next number in the pattern.**

STEP 1   Decide if the numbers increase or decrease.

The numbers decrease.

STEP 2   Find a rule for the pattern.

Since the numbers are decreasing, use subtraction or division.

Try subtraction.

$150 - ? = 135$ → $150 - \mathbf{15} = 135$

$135 - ? = 120$ → $135 - \mathbf{15} = 120$

$120 - ? = 105$ → $120 - \mathbf{15} = 105$

A rule is subtract 15.

STEP 3   Use the rule to find the next number.

$105 - \mathbf{15} = 90$

SOLUTION   **Mateo will have $90 in his account on Friday.**

Patterns can also use geometric figures. Geometric patterns can be repeating or growing patterns. As with a numerical pattern, find the rule to extend the pattern.

EXAMPLE 4

What is the seventeenth figure in this pattern?

STRATEGY   **Find the rule of the pattern. Then extend it.**

STEP 1   Find the rule of the pattern.

The pattern does not grow. To find the rule of a repeating pattern, see which figures repeat. The rule is 1 triangle, 1 rectangle, and 2 circles.

STEP 2   Extend the pattern.

You know the first 8 figures. Make a table for figures 9 through 17.

Figure	9	10	11	12	13	14	15	16	17
Shape	△	▯	◯	◯	△	▯	◯	◯	△

SOLUTION   **The seventeenth figure in this pattern is a triangle.**

EXAMPLE 5

How many dots will be in the next figure?

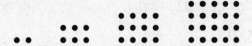

STRATEGY   **Find the rule for the pattern. Then extend the pattern.**

STEP 1   Count the number of dots in each figure.

Figure	1	2	3	4
Number of Dots	2	6	12	20

Since the number of dots grows, it is a growing pattern.

STEP 2   Find the rule.

Figure	1	2	3	4
Pattern	1 × 2	2 × 3	3 × 4	4 × 5

The rule is each figure has 1 more dot in each row and 1 more row than the previous figure.

STEP 3   Use the rule to find how many dots will be in the next figure.

There will be 5 rows of 6 dots in the next figure.

6 × 5 = 30

SOLUTION   **There will be 30 dots in the next figure.**

COACHED EXAMPLE

What is the next number in this pattern?

183, 191, 199, 207, 215, …

THINKING IT THROUGH

Do the numbers in the pattern increase or decrease? _____

Check if the numbers increase by an equal amount each time.

183 + 8 = 191

191 ◯ _____ = _____

199 ◯ _____ = _____

207 ◯ _____ = _____

The rule for this pattern is _____.

Use the rule to find the next number in the pattern.

_____ ◯ _____ = _____

The next number in the pattern is _____.

Lesson Practice

Choose the correct answer.

1. Which pattern follows the rule subtract 7?

   A. 72, 66, 59, 53, 46

   B. 81, 74, 67, 60, 53

   C. 76, 70, 64, 58, 52

   D. 72, 79, 86, 93, 100

2. What is the next number in this pattern?

   6, 14, 22, 30, 38, …

   A. 40

   B. 42

   C. 46

   D. 48

3. Jamal made a pattern with the rule multiply by 3. Which could be Jamal's pattern?

   A. 3, 9, 18, 27

   B. 2, 6, 12, 36

   C. 3, 6, 9, 12

   D. 2, 6, 18, 54

4. Which is the missing number in this pattern?

   13, 22, 31, _____, 49, 58

   A. 39

   B. 40

   C. 41

   D. 42

5. Mr. Collins wrote this pattern on the board.

   96, 48, 24, 12, 6, …

   He asked his class to use the same rule to write a similar pattern beginning with 72. Which is the next number in the new pattern?

   A. 4

   B. 9

   C. 18

   D. 36

6. If this pattern continues, how many dots will be in the next figure?

   A. 10

   B. 12

   C. 16

   D. 24

7. Which figure replaces the question mark in this pattern?

A. ⌐|

C. |⌐

B. ⌐⌐

D. ⌐|

8. Which figure comes next in this pattern?

◇ ◯ □ ◇ ◯ □ ?

A. ◇

C. □

B. ◯

D. ◿

9. What is a rule for this pattern?

4, 8, 16, 32, 64, ...

*Answer*_____

10. Henry receives money from his grandparents every birthday as shown in the table below.

Birthday	Amount
1st	$1
2nd	$4
3rd	$7
4th	$10

If the pattern continues, how much money will Henry receive on his eighth birthday?

*Answer*_____

EXTENDED-RESPONSE QUESTION

11. Kathy wrote the following number pattern in her notebook.

75, 70, 65, 60, 55, 50, ...

Part A   What is the rule for Kathy's pattern?

_____

Part B   If Kathy continues her pattern, what will be the ninth number in her pattern?

_____

25 Input-Output Tables

4.A.5

Getting the Idea

You can use an **input-output table** to show how two groups of numbers are related. Look at the table below.

Set A	Set B
5	10
7	14
9	18
12	24

Each row shows a pair of numbers that are related. For each row, the number in Set B is twice the number in Set A. The rule for the table is multiply by 2.

You can check that the rule is correct by applying it to all the rows in the table.

$5 \times 2 = 10$

$7 \times 2 = 14$

$9 \times 2 = 18$

$12 \times 2 = 24$

The rule works for all the numbers in the table, so it is correct.

EXAMPLE 1

What is the missing number in this input-output table?

Input	Output
10	15
17	22
23	28
32	
41	46

STRATEGY   **Find the rule. Use it to find the missing number.**

STEP 1   The output is greater than the input. See if there is an addition pattern by adding the same value to each input number.

$$10 + 5 = 15$$

$$17 + 5 = 22$$

$$23 + 5 = 28$$

$$41 + 5 = 46$$

The rule is input + 5 = output.

STEP 2   Find the missing number by applying the rule.

$$32 + 5 = 37$$

SOLUTION   **The missing number is 37.**

EXAMPLE 2

Kate is baking apple pies for the school bake sale. The table shows how many apples she will need for different numbers of pies. How many apples will Kate need to be able to bake 8 pies?

Apples for Pies

Number of Pies	Number of Apples
2	8
3	12
4	16
5	20
6	24

STRATEGY   Find the rule.

STEP 1   The numbers increase, so check if this is an addition or a multiplication rule.

Subtract to see if the number increases by the same amount each time.

$8 - 2 = 6$

$12 - 3 = 9$

$16 - 4 = 12$

This is not an addition rule.

STEP 2   Divide the numbers to find the multiplication rule for the table.

$8 \div 2 = 4$

$12 \div 3 = 4$

$16 \div 4 = 4$

$20 \div 5 = 4$

$24 \div 6 = 4$

The rule is multiply by 4.

STEP 3   Apply the rule.

$8 \times 4 = 32$

SOLUTION   **Kate will need 32 apples to bake 8 pies.**

COACHED EXAMPLE

Look at the input-output table.

Input	Output
4	20
7	35
11	55
14	70

What is the output number if the input number is 18?

THINKING IT THROUGH

The output numbers are _____ than the input numbers.

The rule involves _____ or _____.

The numbers _____ _____ increase by the same amount, so this is a _____ rule.

Divide to find a rule.

$20 \div 4 =$ _____

$35 \div 7 =$ _____

$55 \div 11 =$ _____

$70 \div 14 =$ _____

A rule for the table is_____.

Apply the rule. $18 \times$ _____ $=$ _____

When the input number is 18, the output number is _____.

Lesson Practice

Choose the correct answer.

1. What is the missing number in this input-output table?

Input	Output
3	9
5	15
8	24
11	
14	42

   A. 27

   B. 30

   C. 33

   D. 39

2. What is a rule of this input-output table?

Input	Output
5	10
8	16
12	24
15	30

   A. input × 2 = output

   B. input + 5 = output

   C. input × 5 = output

   D. input + 5 = output

3. What will be the value of the output if the input is 72 in the table below?

Input	Output
12	2
24	4
36	6
48	8

   A. 9

   B. 10

   C. 11

   D. 12

4. Which describes a rule of this input-output table?

Input	Output
3	48
7	112
12	192
18	288

   A. input + 45 = output

   B. input + 270 = output

   C. input × 14 = output

   D. input × 16 = output

5. Write a rule for the following input-output table.

Input	Output
12	19
19	26
31	38
42	49

*Answer*_____

6. What is the output number when the input number is 62 in the table below?

Input	Output
13	4
24	15
36	27
48	39

*Answer*_____

EXTENDED-RESPONSE QUESTION

7. Bryn needs to make 150 string bracelets for the school fair. The table below shows the total number of bracelets that Bryn has made by the end of days 3 through 6.

Day	Number of String Bracelets
3	15
4	20
5	25
6	30

Part A Using the pattern in the table above, how many string bracelets will Bryn have made by the end of Day 9?

_____

Part B If the pattern in the table continues, how many days will it take Bryn to make the 150 string bracelets that she needs for the fair? Explain how you know your answer is correct.

_____

_____

_____

2 Review

1 Which number makes this sentence true?

2,742 < _____

A 2,247

B 2,422

C 2,428

D 2,760

2 What is the rule of this input-output table?

Input	Output
6	11
8	13
12	17
15	20

A Multiply by 4.

B Multiply by 2.

C Subtract 5.

D Add 5.

3 Rich writes the pattern below.

8, 17, 26, 35, _____

What is the next number in Rich's pattern?

A 44   C 46

B 45   D 47

4 Emma baked 12 cookies to give to 6 friends. Which open sentence can be used to find how many cookies each friend will get?

A 12 ÷ 6 = ____

B 12 × 6 = ____

C 12 − 6 = ____

D 12 + 6 = ____

5 Which number makes this sentence true?

3,629 > _____

A 6,329   C 3,692

B 3,962   D 3,269

6 What is the rule of this input-output table?

Input	Output
3	12
5	20
8	32
11	44

A Multiply by 4.

B Multiply by 5.

C Divide by 4.

D Add 9.

7 The table shows the amount of money that Carla earns for walking dogs.

CARLA'S DOG WALKING

Number of Walks	Amount Earned
2	$10
3	$15
4	$20
5	$25

If the pattern in the table continues, how much money will Carla earn if she walks 8 dogs?

A $32

B $40

C $48

D $56

8 Carlos has a box of 64 crayons. He gives his sister 20 of the crayons. Which open sentence shows how to find how many crayons Carlos has now?

A $64 + 20 =$ _____

B $64 \times 20 =$ _____

C $64 - 20 =$ _____

D $64 \div 20 =$ _____

9 Which number makes this sentence true?

$$5,132 < \text{\_\_\_\_\_} < 5,164$$

A 5,123   C 5,213

B 5,148   D 5,416

10 Which input-output table follows the rule below?

$$\text{Input} - 6 = \text{Output}$$

A

Input	Output
12	2
18	3
24	4
30	5

B

Input	Output
10	2
15	3
20	4
25	5

C

Input	Output
8	2
12	6
15	9
21	15

D

Input	Output
3	9
5	11
7	13
10	16

11 Tommy drew this pattern of small squares.

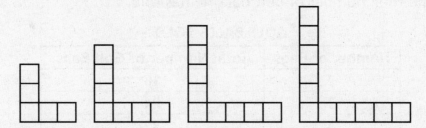

If the pattern continues, how many small squares are in the next figure?

A 12

B 13

C 14

D 15

12 Melissa is 8 years younger than Francine. Francine is 14 years old. Which number sentence could be used to find Melissa's age?

A 14 + 8 = ☐

B 14 ÷ 8 = ☐

C 14 − 8 = ☐

D 14 × 8 = ☐

13 Vance is putting used golf balls in bags to sell to a driving range. The table shows the total number of golf balls he has sold.

GOLF BALLS SOLD

Number of Bags	Total Number of Golf Balls
2	16
3	24
4	32
5	40

If the pattern in the table continues, how many golf balls will Vance have sold if he sells 8 bags?

Show your work.

Answer _____ golf balls

14 Nick is 6 years older than Juliana. If Nick is 28 years old, write an open sentence that can be used to find Juliana's age, in years.

Answer _____

15 Roger writes the number pattern below.

4, 12, 20, 28, 36, _____, _____, _____

Part A

What is the rule of Roger's pattern?

Rule _____

Part B

What are the next three numbers in Roger's pattern?

Answer _____

Part C

On the lines below, explain if 72 is part of Roger's pattern.

_____

_____

_____

_____

STRAND

3 Geometry

			NYS Math Indicators
Lesson 26	Lines	168	4.G.6**
Lesson 27	Angles	175	4.G.7**, 4.G.8**
Lesson 28	Two-Dimensional Figures	181	4.G.1, 4.G.2
Lesson 29	Perimeter and Area	187	4.G.3, 4.G.4
Lesson 30	Three-Dimensional Figures	193	4.G.5
Strand 3 Review		197	

** Grade 4 May–June Indicators

26 Lines

4.G.6

Getting the Idea

A **point** is a particular place or location.

A **line** is a straight path that goes in two directions without end. This line can be written as $\overleftrightarrow{ST}$ or $\overleftrightarrow{TS}$.

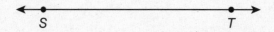

A **ray** is part of a line that has one **endpoint** and goes in the other direction without end. A ray is named by its endpoint first. This ray is named $\overrightarrow{YZ}$.

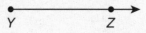

A **line segment** is part of a line that has two endpoints. This line segment can be named as $\overline{MN}$ or $\overline{NM}$.

EXAMPLE 1

Name the figure.

STRATEGY   Look at the figure to find identifying features.

STEP 1   Identify the figure.

The figure shows a straight path going on without end in both directions. The figure is a line.

STEP 2   Identify points on the line.

The points B and C are located on the line.

STEP 3   Name the line.

Use the points on the line to name it.

SOLUTION   **The name of the figure is line BC or line CB.**

EXAMPLE 2

Draw $\overrightarrow{DE}$.

STRATEGY   **Identify the figure and then draw the figure.**

STEP 1   Identify the figure.

The figure is a ray.

STEP 2   Draw and label the endpoint.

The endpoint is the first point listed in the name.

D is the endpoint.

•
D

STEP 3   Draw the ray.

Draw an arrow pointing away from point D.

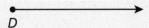
D

STEP 4   Draw and label the second point on the ray.

The other point on the ray is point E.

D   E

SOLUTION   **Ray DE is shown in Step 4.**

Pairs of lines or line segments can be identified as parallel, intersecting, or perpendicular.

Parallel lines are lines that remain the same distance apart and never meet.

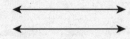

Intersecting lines are lines that cross.

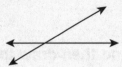

Perpendicular lines are intersecting lines that cross to form 4 **right angles.**

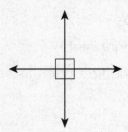

EXAMPLE 3

Which street is parallel to 2nd Avenue?

STRATEGY   Use the definition of parallel lines.

Parallel lines are lines that remain the same distance apart and never meet. First Avenue and 2nd Avenue remain the same distance apart. They do not meet.

SOLUTION   First Avenue is parallel to 2nd Avenue.

EXAMPLE 4

Which figure includes perpendicular line segments?

  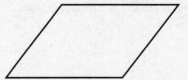

STRATEGY   **Look for right angles.**

STEP 1   Look at the first figure.

The line segments do not meet at right angles, so none of the line segments are perpendicular.

STEP 2   Look at the second figure.

The line segments meet at right angles, so the figure includes line segments that are perpendicular.

STEP 3   Look at the third figure.

The line segments do not meet at right angles, so none of the line segments are perpendicular.

SOLUTION   **The second figure includes line segments that are perpendicular.**

COACHED EXAMPLE

Figure *ABCD* is shown below.

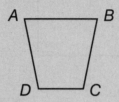

What kind of lines are $\overline{AB}$ and $\overline{CD}$?

THINKING IT THROUGH

Do $\overline{AB}$ and $\overline{CD}$ meet or cross? _____

Do $\overline{AB}$ and $\overline{CD}$ meet or cross at a right angle? _____

Line segments *AB* and *CD* are _____.

Lesson Practice

Choose the correct answer.

1. Which figure shows line *XY*?

   A. X———————Y

   B. ←X———————Y→

   C. X———————Y→

   D. Y———————X→

2. Which figure is shown below?

   A   B

   A. line segment *AB*
   B. line *AB*
   C. ray *AB*
   D. ray *BA*

3. Which does **not** name a ray in the figure shown below?

   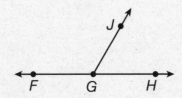

   A. $\overrightarrow{GF}$
   B. $\overleftrightarrow{FH}$
   C. $\overrightarrow{GJ}$
   D. $\overrightarrow{GH}$

4. Which pair of lines are perpendicular?

   A.
   B.
   C.
   D.

5. Which describes a figure with two endpoints that is part of a line?

   A. $\overline{LM}$ C. $\overrightarrow{LM}$
   B. $\overleftrightarrow{LM}$ D. $\overrightarrow{ML}$

6. Which describes the line segments in this figure?

   A. parallel and intersecting, but not perpendicular
   B. parallel only
   C. intersecting only
   D. parallel and perpendicular

Use the diagram for questions 7 and 8.

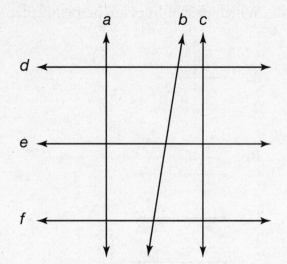

7. Which pair of lines are intersecting, but **not** perpendicular?

   A. line *a* and line *c*

   B. line *b* and line *f*

   C. line *d* and line *e*

   D. line *a* and line *f*

8. Which describes lines *a* and *c*?

   A. perpendicular

   B. intersecting but not perpendicular

   C. neither parallel nor intersecting

   D. parallel

9. Draw parallel lines *OP* and *JK*.

10. Draw intersecting lines *PQ* and *TU*.

27 Angles

4.G.7, 4.G.8

Getting the Idea

An **angle** is formed by two rays that have the same endpoint. The common endpoint is the **vertex** of the angle. When naming an angle, the vertex is the middle letter. An angle can also be named by its vertex.

The angle below can be named ∠B, ∠ABC, or ∠CBA.

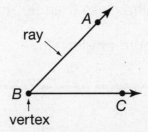

There are four types of angles. Angles are measured in **degrees** (°).

A **right angle** is an angle that forms a square corner. It measures exactly 90°.

An **acute angle** forms an angle less than a right angle. It measures more than 0° but less than 90°.

An **obtuse angle** forms an angle greater than a right angle. It measures more than 90° but less than 180°.

A **straight angle** measures 180°.

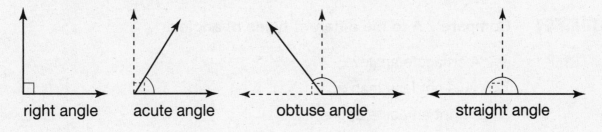

right angle   acute angle   obtuse angle   straight angle

EXAMPLE 1

What kind of angle is shown?

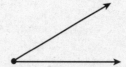

STRATEGY **Compare the angle to a right angle.**

STEP 1 Decide if the angle is a right angle.

The angle is not a right angle.

STEP 2 Compare the angle measure to a right, or 90°, angle.

The angle measures less than 90° but more than 0°.

SOLUTION **The angle is an acute angle.**

EXAMPLE 2

The hands on the clock form an angle. What type of angle is ∠A?

STRATEGY **Compare ∠A to the different types of angles.**

STEP 1 Is ∠A an acute angle?

∠A is not less than 90°.

It is not an acute angle.

STEP 2 Is ∠A a right angle?

∠A is not exactly 90° and does not form a square corner.

It is not a right angle.

STEP 3   Is ∠A an obtuse angle?

∠A is greater than 90° and is less than 180°.

It is an obtuse angle.

SOLUTION   **∠A is an obtuse angle.**

EXAMPLE 3

Draw an angle named ∠RST.

STRATEGY   **Use a ruler or straightedge to draw the angle.**

STEP 1   Draw the vertex of the angle.

The vertex of angle RST is the middle letter.

Draw endpoint S.

•
S

STEP 2   Draw ray ST.

Use a straightedge to draw ray ST starting at endpoint S.
Label point T on the ray.

STEP 3   Draw the angle's other ray.

Use a straightedge to draw ray SR starting at endpoint S.

Label point R on the ray.

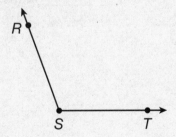

SOLUTION   **∠RST is shown in Step 3.**

COACHED EXAMPLE

Classify the angles of this triangle.

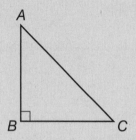

THINKING IT THROUGH

Look at $\angle A$.

Compare it to a right angle.

$\angle A$ is _____ than a right angle.

$\angle A$ is a(n) _____ angle.

Look at $\angle B$.

The symbol at $\angle B$ means it is a(n) _____ angle.

Look at $\angle C$.

Compare it to a right angle.

$\angle C$ is _____ than a right angle.

$\angle C$ is an _____ angle.

$\angle A$ is an _____ angle, $\angle B$ is a _____ angle, and $\angle C$ is an _____ angle.

Lesson Practice

Choose the correct answer.

1. Which is an obtuse angle?

   A.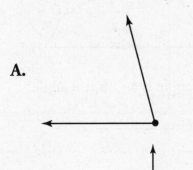

   B.

   C.

   D.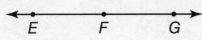

2. Which type of angle is shown below?

   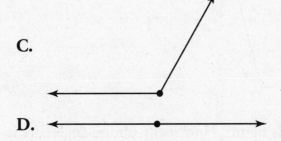

   A. straight
   B. right
   C. acute
   D. obtuse

3. Which is **not** a way to name this angle?

   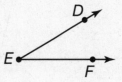

   A. ∠DEF   C. ∠E
   B. ∠FDE   D. ∠FED

4. Wendy was driving with her mother when they passed a yield sign. Which type of angles does the sign have?

   A. straight   C. right
   B. obtuse   D. acute

5. Which definition best describes a right angle?

   A. The angle measurement equals 90°.

   B. The angle has a measurement greater than 90° but less than 180°.

   C. The angle measurement equals 180°.

   D. The angle measurement is between 0° and 90°.

6. Which type of angle is shown below?

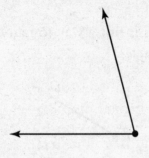

Answer _____

7. Jessica drew the figure below.

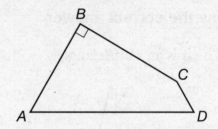

Which angle is an obtuse angle?

Answer _____

EXTENDED-RESPONSE QUESTION

8. Look at figure *MNOP* below.

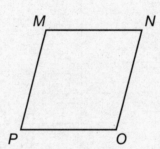

Part A How many acute angles are in this figure? How many obtuse angles are in this figure?

_____

Part B Name the acute angles. Name the obtuse angles.

acute angles _____

obtuse angles _____

28 Two-Dimensional Figures

4.G.1, 4.G.2

Getting the Idea

A **polygon** is a closed **two-dimensional figure** with straight sides. Polygons are classified, or sorted, by the number of sides and angles. Each side of a polygon is a line segment. The line segments meet at points that form the corners of the polygon.

triangle	quadrilateral	pentagon	hexagon	octagon
3 sides 3 angles	4 sides 4 angles	5 sides 5 angles	6 sides 6 angles	8 sides 8 angles

A **circle** is a closed two-dimensional figure in which all points are an equal distance from the **center**. A circle is not a polygon.

EXAMPLE 1

What is the name of this polygon?

STRATEGY **Count the number of sides.**

There are 5 line segments.

Each of the line segments is a side.

SOLUTION **The polygon is a pentagon.**

A **quadrilateral** is a polygon with 4 sides and 4 angles. Quadrilaterals are classified by the measures of their sides and angles. Here are some quadrilaterals:

parallelogram	rectangle	square	rhombus	trapezoid
both pairs of opposite sides are parallel	parallelogram with 4 right angles	parallelogram with 4 right angles with all sides the same length	parallelogram with all sides the same length	exactly 1 pair of parallel sides

EXAMPLE 2

What is the name of this figure? Be as specific as possible.

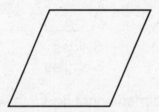

STRATEGY   **Identify the relationship between the sides.**

STEP 1   Is it a quadrilateral?

     Yes, it has 4 straight sides.

STEP 2   Is it a parallelogram?

     Yes, it has 2 pairs of parallel sides.

STEP 3   Are the angles right angles?

     No, none of the angles are right angles.

STEP 4   Are the sides all the same length?

     Yes, the sides are all the same length.

     A rhombus is a parallelogram with all sides the same length.

SOLUTION   **The quadrilateral is a rhombus.**

EXAMPLE 3

Draw the figure described below. Then identify the figure. Be as specific as possible.

This figure has 4 line segments, or sides.

This figure has 4 right angles.

All 4 sides are the same length.

STRATEGY   **Draw the figure using the description. Then identify the figure.**

STEP 1   Draw the figure.

STEP 2   How many sides and angles does the figure have?
It has 4 sides and 4 angles.
The figure is a quadrilateral.

STEP 3   What other special features does the figure have?
It has 4 right angles and all the sides are the same length.
A square matches this description.

SOLUTION   **The figure is a square.**

COACHED EXAMPLE

Mr. Gilroy is drawing a hexagon on the board. He drew the 3 line segments shown below.

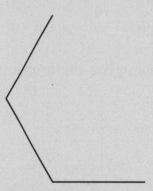

Mr. Gilroy asks a student to complete the hexagon. How many more line segments must the student draw to complete the hexagon?

THINKING IT THROUGH

How many line segments does a hexagon have? _____

How many line segments has Mr. Gilroy drawn so far? _____

How many line segments are missing?

_____ −3 = _____ missing line segments

The student must draw _____ more lines segments to complete the hexagon.

Lesson Practice

Choose the correct answer.

1. Which best describes this polygon?

   A. rectangle

   B. rhombus

   C. square

   D. trapezoid

2. Which polygon is shown below?

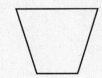

   A. hexagon

   B. octagon

   C. pentagon

   D. triangle

3. Which is **not** a polygon?

   A.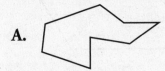

   B.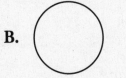

   C.

   D.

4. What is the name of the polygon below?

  A. triangle

  B. quadrilateral

  C. pentagon

  D. octagon

5. Mike drew a polygon with 3 sides and 3 angles. Which could be Mike's figure?

  A. circle

  B. square

  C. triangle

  D. pentagon

6. How many angles does an octagon have?

  A. 6

  B. 7

  C. 8

  D. 9

7. Frank wants to draw a pentagon. How many line segments should he draw to make this polygon?

  Answer_____

8. Name a polygon that has 4 sides and 4 angles.

  Answer _____

29 Perimeter and Area

 4.G.3, 4.G.4

Getting the Idea

Perimeter is the distance around a figure. To find the perimeter of a figure, add the measures of all the sides.

EXAMPLE 1

What is the perimeter, in centimeters, of this rectangle?

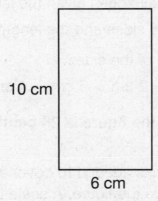

10 cm

6 cm

STRATEGY **Add the lengths of the sides.**

STEP 1 Determine if you have been given the length of every side.

A rectangle has 4 sides, with opposite sides having equal lengths.

So both of the two longer sides measure 10 cm each and both of the two shorter sides measure 6 cm each.

STEP 2 Add the measures of the sides.

10 cm + 10 cm + 6 cm + 6 cm = 32 cm

SOLUTION **The perimeter of the rectangle is 32 centimeters.**

EXAMPLE 2

What is the perimeter, in centimeters, of this figure?

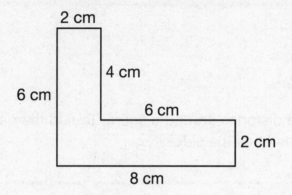

STRATEGY   **Add the lengths of the sides.**

STEP 1   Determine if you have been given the lengths of every side.

The figure has 6 sides and the length of each side is given.

STEP 2   Add the measures of the sides.

6 cm + 8 cm + 2 cm + 6 cm + 4 cm + 2 cm = 28 cm

SOLUTION   **The perimeter of the figure is 28 centimeters.**

Area is the number of square units needed to cover a figure. To find the area, count the number of square units inside the figure. A scale tells what each square unit represents. Some examples of units used for area are square inches, square feet, and square centimeters.

EXAMPLE 3

What is the area, in square meters, of this rectangle?

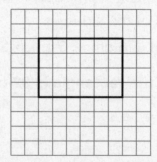

Key: □ = 1 square meter

STRATEGY   Count the number of square units inside the figure.

STEP 1   Count the square units.

There are 24 square units inside the rectangle.

STEP 2   Use the scale to find what each square unit represents.

Each square unit is 1 square meter.

So 24 square units = 24 square meters.

SOLUTION   **The area of the rectangle is 24 square meters.**

COACHED EXAMPLE

Below is a diagram of Leslie's backyard. Find the perimeter and area of Leslie's backyard.

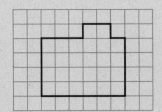

Key: ⊢ = 1 unit
☐ = 1 square unit

THINKING IT THROUGH

To find the perimeter, _____ the measures of all the sides.

_____ + _____ + _____ + _____ + _____ + _____ + _____ +
_____ = _____

The perimeter is _____ units.

To find the area, _____ the number of square units inside the figure.

There are _____ square units inside the figure.

The area is _____ square units.

The perimeter of Leslie's backyard is _____ units and the area is _____ square units.

Lesson Practice

Choose the correct answer.

1. What is the perimeter, in millimeters, of this square?

   8 mm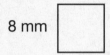

   A. 16 mm

   B. 32 mm

   C. 48 mm

   D. 64 mm

2. What is the perimeter, in inches, of this rectangle?

   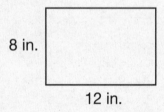

   8 in.

   12 in.

   A. 96 in.

   B. 40 in.

   C. 32 in.

   D. 28 in.

3. A drawing of Clarissa's school playground is shown below. What is the perimeter of Clarissa's school playground?

   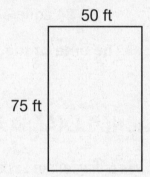

   50 ft

   75 ft

   A. 125 feet

   B. 175 feet

   C. 200 feet

   D. 250 feet

4. What is the area, in square units, of this rectangle?

   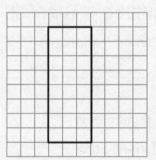

   Key: ☐ = 1 square unit

   A. 24 square units

   B. 22 square units

   C. 12 square units

   D. 11 square units

5. Lisa made a drawing of her kitchen. What is the area, in square feet, of Lisa's kitchen?

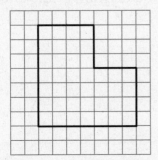

Key: ☐ = 1 square foot

A. 12 square feet

B. 28 square feet

C. 40 square feet

D. 80 square feet

6. What is the area, in square inches, of the figure below?

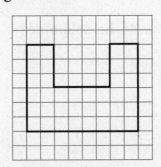

Key: ☐ = 1 square inch

A. 48 square inches

B. 36 square inches

C. 28 square inches

D. 24 square inches

7. Trevor drew the following diagram of his bedroom. What is the perimeter, in feet, of Trevor's bedroom?

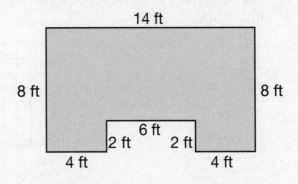

Answer _____

8. An architect is designing an office for his client. Below is a diagram of the office space. What is the area, in square yards, of the office?

Office Space

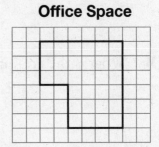

Key: ☐ = 1 square yard

Answer _____

EXTENDED-RESPONSE QUESTION

9. Claudia wants to build a miniature dollhouse. She drew a diagram of the dollhouse floor below.

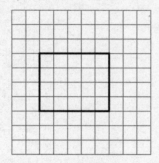

Key: ⊢ = 1 inch

☐ = 1 square inch

Part A What is the perimeter, in inches, of the floor of the dollhouse?

Part B What is the area, in square inches, of the floor of the dollhouse?

30 Three-Dimensional Figures

4.G.5

Getting the Idea

Three-dimensional figures are also known as solid figures. They have length, width, and height. These figures can be sorted by the number of faces, edges, and vertices.

A **face** is the flat surface on a solid figure.

An **edge** is formed at the line segment where two faces meet.

A **vertex** (plural: **vertices**) is the point where three or more edges meet.

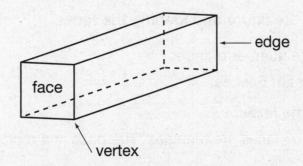

These three-dimensional figures have only flat faces. Prisms and pyramids are named by their **bases**. A base is the face that is used to name a solid figure. Count the number of faces, edges, and vertices of each figure.

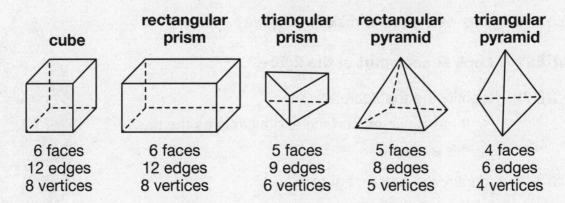

cube	rectangular prism	triangular prism	rectangular pyramid	triangular pyramid
6 faces	6 faces	5 faces	5 faces	4 faces
12 edges	12 edges	9 edges	8 edges	6 edges
8 vertices	8 vertices	6 vertices	5 vertices	4 vertices

The three-dimensional figures shown below have curved surfaces.

sphere	cone	cylinder
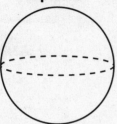		
1 curved surface	1 circular base 1 curved surface 1 vertex	2 congruent circular bases 1 curved surface

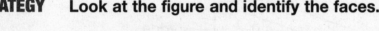

EXAMPLE 1

Which polygons form the faces of a rectangular pyramid?

STRATEGY **Look at the figure and identify the faces.**

STEP 1 Count the number of faces.

There are 5 faces.

STEP 2 Identify the faces.

4 of the faces are triangles. The base is a rectangle.

SOLUTION **The faces of a rectangular pyramid are 4 triangles and 1 rectangle.**

EXAMPLE 2

Which two figures were used to make this figure?

STRATEGY **Look at each part of the figure.**

STEP 1 Identify the top figure.

It has a curved surface and a circle as the base.

It is a cone.

STEP 2 Identify the bottom figure.

It has a curved surface and 2 faces that are circles.

It is a cylinder.

SOLUTION **The figures are a cone and a cylinder.**

COACHED EXAMPLE

Which figure has the same number of vertices as faces?

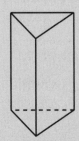
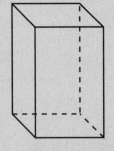
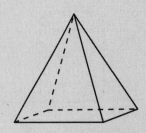

triangular prism   **rectangular prism**   **square pyramid**

THINKING IT THROUGH

How many faces does the triangular prism have? _____

How many vertices does the triangular prism have? _____

How many faces does the rectangular prism have? _____

How many vertices does the rectangular prism have? _____

How many faces does the square pyramid have? _____

How many vertices does the square pyramid have? _____

The _____ _____ has the same
number of vertices as faces.

Lesson Practice

Choose the correct answer.

1. Name the solid figure below.

   A. sphere   C. cylinder

   B. cube   D. cone

2. Mrs. Patterson asked Kelly to name the figure that has only 1 square face. What should Kelly give as her answer?

   A. cylinder

   B. square pyramid

   C. cube

   D. cone

3. How many vertices does the figure below have?

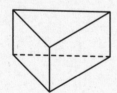

   A. 3

   B. 5

   C. 6

   D. 9

4. Which figure has 4 triangular faces?

   A. triangular prism

   B. cube

   C. rectangular prism

   D. triangular pyramid

5. Which three-dimensional figure does **not** have a curved surface?

   A. cone

   B. cylinder

   C. cube

   D. sphere

6. What is the name of the solid figure below?

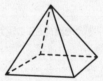

   Answer _____

3 Review

1 What is the perimeter of this figure?

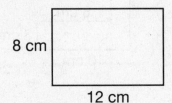

8 cm

12 cm

   A   20 centimeters

   B   40 centimeters

   C   48 centimeters

   D   96 centimeters

2 What is the name of the shape below?

   A   quadrilateral

   B   pentagon

   C   hexagon

   D   octagon

3 How many faces does this triangular prism have?

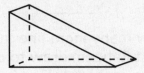

   A   3

   B   4

   C   5

   D   6

4 What is the area of this rectangle?

KEY
☐ = 1 square unit

   A   22 square units

   B   24 square units

   C   30 square units

   D   36 square units

5 Raheem's homework includes drawing a quadrilateral. Raheem draws the 2 line segments shown below before class ends.

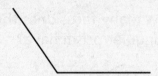

How many **more** line segments must Raheem draw to complete the quadrilateral?

A 1

B 2

C 3

D 4

6 How can the pair of lines shown below be classified?

A intersecting and perpendicular

B intersecting but not perpendicular

C perpendicular but not intersecting

D parallel

Use the shape for questions 7 and 8.

Lucille drew the following shape.

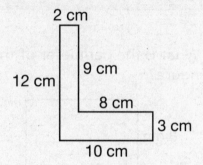

7 What is the name of this shape?

A quadrilateral

B pentagon

C hexagon

D octagon

8 What is the perimeter of this shape?

A 44 centimeters

B 42 centimeters

C 40 centimeters

D 38 centimeters

9 How many edges does a cube have?

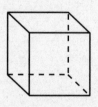

A 6　　　　C 10

B 8　　　　D 12

10 What is the area of this rectangle?

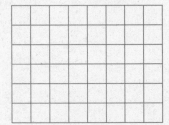

KEY
☐ = 1 square unit

A   48 square units

B   42 square units

C   40 square units

D   28 square units

11 What is the perimeter of this shape?

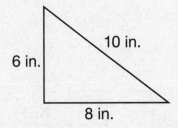

A   22 inches   C   26 inches

B   24 inches   D   28 inches

12 How can this angle be classified?

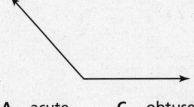

A   acute   C   obtuse

B   right   D   straight

13 Which is a ray in the angle shown below?

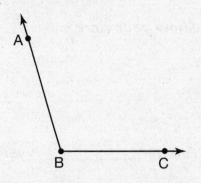

A   ray AB   C   ray AC

B   ray CB   D   ray BA

14 Which shape has 8 sides and 8 angles?

A   quadrilateral

B   pentagon

C   hexagon

D   octagon

15 What is the area of this rectangle?

KEY
☐ = 1 square unit

A   35 square units

B   30 square units

C   24 square units

D   12 square units

16 Leo has a rectangular backyard that is 42 yards long and 36 yards wide. What is the perimeter of Leo's backyard?

Show your work.

Answer _____ yards

17 In the space below, draw a pair of lines that are perpendicular.

18 Which two rectangles have the same area?

KEY
☐ = 1 square unit

Show your work.

Answer _____

What is the area, in square units, of rectangle A?

Answer _____ square units

19   Ana drew this figure.

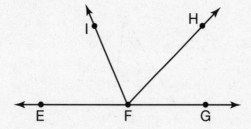

Part A

Name three acute angles in the figure.

Answer _____

Part B

Name two obtuse angles in the figure.

Answer _____

Part C

Name one straight angle in the figure.

Answer _____

20 Charlotte received a gift box in the shape of a square pyramid as shown below.

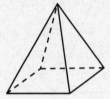

Part A

Describe the shapes of the faces of a square pyramid.

Answer _____

Part B

How many vertices does a square pyramid have?

Answer _____ vertices

Part C

How many edges does a square pyramid have?

Answer _____ edges

STRAND

4 Measurement

			NYS Math Indicators
Lesson 31	Customary Units of Length	204	4.M.1, 4.M.2, 4.M.3
Lesson 32	Metric Units of Length	211	4.M.1, 4.M.2
Lesson 33	Metric Units of Mass	217	4.M.4, 4.M.5
Lesson 34	Metric Units of Capacity	222	4.M.6, 4.M.7
Lesson 35	Make Change	225	4.M.8
Lesson 36	Elapsed Time	231	4.M.9, 4.M.10
	Strand 4 Review	236	

31 Customary Units of Length

4.M.1, 4.M.2, 4.M.3

Getting the Idea

When you want to know how long or tall something is, you measure its **length**. The table below shows the units for length in the **customary system**.

Customary Units of Length
1 foot (ft) = 12 inches (in.)
1 yard (yd) = 3 feet
1 yard = 36 inches
1 mile (mi) = 5,280 feet

This eraser is about 1 **inch** long.

A hardcover book is about 1 **foot** long.

A baseball bat is about 1 **yard** long.

1 **mile** is the distance an adult can walk in about 20 minutes.

To measure the length of small objects, you can use an inch ruler. To measure greater lengths, use a yardstick or a tape measure. An odometer can be used to measure distance in miles.

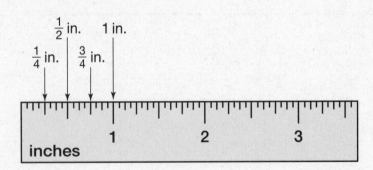

EXAMPLE 1

What is the length of this pen, to the nearest $\frac{1}{2}$ inch?

STRATEGY   Use an inch ruler.

STEP 1   Line up the left end of the ruler with the left end of the pen.

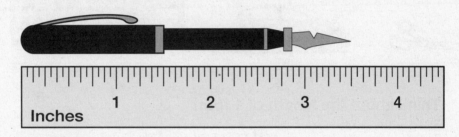

STEP 2   Look at the right end of the pen.

It is halfway between 3 and 4 inches.

The pen is closest to $3\frac{1}{2}$ inches.

SOLUTION   To the nearest $\frac{1}{2}$ inch, the pen is $3\frac{1}{2}$ inches long.

When converting units of measurement, remember:

To convert a larger unit to a smaller unit, multiply.

To convert a smaller unit to a larger unit, divide.

EXAMPLE 2

Mr. Conroy is 6 feet tall. How tall is Mr. Conroy in inches?

STRATEGY   Multiply to convert a larger unit to a smaller unit.

STEP 1   How many inches are in 1 foot?
There are 12 inches in 1 foot.

STEP 2   Multiply 6 by 12.
$6 \times 12 = 72$

SOLUTION   Mr. Conroy is 72 inches tall.

EXAMPLE 3

Which real object could be about 150 feet long?

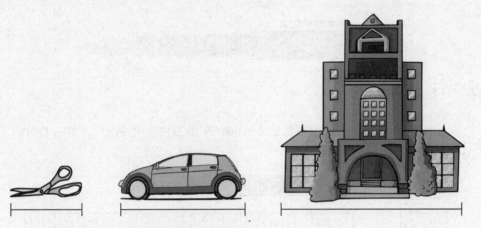

STRATEGY   **Think about the length of 1 foot.**

STEP 1   Think about how long 150 feet will be.

> 1 foot is about the length of this book.
>
> Imagine lining up 150 books. That is the length of about 150 feet.

STEP 2   Look at the objects.

> A pair of scissors is much shorter than 150 feet.
>
> A car is shorter than 150 feet.
>
> A building could be about 150 feet long.

SOLUTION   **The building could be about 150 feet long.**

EXAMPLE 4

Rhonda drove from her home in Syracuse to her grandmother's house in Rochester. Rhonda said the distance is 90. Did Rhonda mean to say 90 inches, 90 feet, or 90 miles?

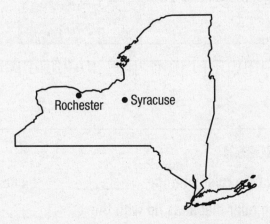

STRATEGY   **Think about the customary units of length.**

STEP 1   Think about the units. Eliminate those that do not make sense.

This book is about 1 foot long.

An inch is smaller than a foot.

The units are either yards or miles.

STEP 2   Compare yards and miles.

90 yards is about the length of a football field.

The distance between Syracuse and Rochester is much longer than a football field.

SOLUTION   **Rhonda meant to say 90 miles.**

COACHED EXAMPLE

What is the length of the ribbon to the nearest $\frac{1}{4}$ inch?

Inches

1   2   3   4   5

THINKING IT THROUGH

Line up the left end of the ruler with the _____ end of the ribbon.

Read the mark on the ruler that lines up with the _____ end of the ribbon.

The right end of the ribbon is between the labeled inch marks _____ and _____ on the ruler.

The end of the ribbon is closest to the _____ inch mark between 4 and 5.

To the nearest $\frac{1}{4}$ inch, the ribbon is _____ inches long.

Lesson Practice

Choose the correct answer.

1. Tyrell wants to measure his thumb. Which unit of measure is best for Tyrell to use?

    A. inch

    B. foot

    C. yard

    D. mile

2. Which is the most reasonable measure of the height of a ceiling?

    A. 10 miles

    B. 10 yards

    C. 10 feet

    D. 10 inches

3. Lisa is 4 feet 8 inches tall. What is Lisa's height in inches?

    A. 40 inches

    B. 48 inches

    C. 52 inches

    D. 56 inches

4. Which measure is equal to 20 feet?

    A. 6 yards 2 feet

    B. 10 yards

    C. 200 inches

    D. 300 inches

Use an inch ruler for questions 5 and 6.

5. To the nearest $\frac{1}{2}$ inch, what is the length of this rectangle?

    A. 2 inches

    B. $2\frac{1}{2}$ inches

    C. 3 inches

    D. $3\frac{1}{2}$ inches

6. To the nearest $\frac{1}{4}$ inch, what is the length of this craft stick?

    A. 1 inch

    B. $1\frac{1}{4}$ inches

    C. $1\frac{3}{4}$ inches

    D. $2\frac{1}{4}$ inches

7. The bases on a softball diamond are 60 feet apart. How many yards apart are the bases?

Answer _____

8. Lilly wants to measure the length of her hair. Which customary unit of length would be best for Lilly to use?

Answer _____

EXTENDED-RESPONSE QUESTION

9. Yan wants to buy new shoelaces for his sneakers. He needs to measure a shoelace from his sneaker to know what length shoelaces to buy.

Part A What customary unit of length would be best for Yan to use to measure his shoelace?

_____

Part B Yan buys shoelaces that are 2 feet 3 inches long each. How long, in inches, is each shoelace?

Show your work.

_____

32 Metric Units of Length

 4.M.1, 4.M.2

Getting the Idea

The table below shows the units of length in the **metric system**.

Metric Units of Length
1 centimeter (cm) = 10 millimeters (mm)
1 meter (m) = 100 centimeters
1 meter = 1,000 millimeters
1 kilometer (km) = 1,000 meters

1 **millimeter** is about the thickness of a penny.

This line is about 1 **centimeter** long.

1 **meter** is a little longer than a yard. It is a little longer than a baseball bat.

1 **kilometer** is the distance an adult can walk in about 10 minutes.

To measure the length of small objects, you can use a centimeter ruler. To measure greater lengths, use a meter stick or a tape measure. Distances in kilometers can be measured using an odometer.

EXAMPLE 1

What is the length of this crayon, to the nearest centimeter?

STRATEGY Use a centimeter ruler to measure the length of the crayon.

STEP 1   Line up the left end of the ruler with the left end of the crayon.

STEP 2   Look at the right end of the crayon.

The crayon is closest to the 5-centimeter mark.

SOLUTION   **To the nearest centimeter, the crayon is 5 centimeters long.**

EXAMPLE 2

Which real object could be about 20 millimeters long?

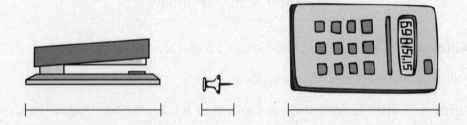

STRATEGY   **Use referents to determine which real object could about 20 millimeters long.**

STEP 1   Think about the length of 1 millimeter.

1 millimeter is about the thickness of a penny.

STEP 2   Think about 20 millimeters.

20 millimeters will be about the height of 15 or 20 stacked pennies.

STEP 3   Look at the objects.

A stapler is longer than 20 millimeters.

A calculator is about the length of a stapler, so it is longer than 20 millimeters.

A pushpin could be about 20 millimeters long.

SOLUTION   **The pushpin could be about 20 millimeters long.**

EXAMPLE 3

Rick said that his bedroom is 5 units long. Which metric unit did Rick mean to say?

STRATEGY   **Think about the metric units of length.**

STEP 1   Think about the units and eliminate those that do not make sense.

This book is about 25 centimeters long, so centimeters and millimeters can be eliminated.

STEP 2   Think about meters and kilometers.

5 meters is about the length of a car.

It would take about 50 minutes to walk 3 kilometers.

SOLUTION   **Rick meant to say 5 meters.**

COACHED EXAMPLE

What is the length of this feather, to the nearest centimeter?

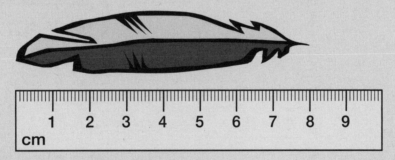

THINKING IT THROUGH

Line up the left end of the ruler with the _____ end of the feather.

Read the mark on the ruler that lines up with the _____ end of the feather.

The right end of the feather is closest to the _____ mark.

To the nearest centimeter, the feather is _____ centimeters long.

Lesson Practice

Choose the correct answer.

1. Which unit is best to measure the length of a playground?

   A. millimeter

   B. centimeter

   C. meter

   D. kilometer

2. Levon wants to measure the length of his foot. Which unit of measure is best for Levon to use?

   A. millimeter

   B. centimeter

   C. meter

   D. kilometer

Use a centimeter ruler for questions 3 and 4.

3. To the nearest centimeter, what is the length of the string?

   A. 4 cm

   B. 5 cm

   C. 6 cm

   D. 8 cm

4. To the nearest centimeter, what is the length of the carrot?

   A. 3 cm

   B. 4 cm

   C. 5 cm

   D. 6 cm

5. The Walter family is driving from Utica, NY to Seneca, NY to visit relatives. Which metric unit is best to measure the distance they need to drive?

    A. kilometer

    B. meter

    C. centimeter

    D. millimeter

6. Serena said that her arm is 50 units long. Which metric unit did Serena mean to say?

    A. millimeters

    B. centimeters

    C. meters

    D. kilometers

7. Which tool is best for measuring the length of a room?

    A. measuring cup

    B. thermometer

    C. balance scale

    D. tape measure

8. Which unit is best to measure the length of a school bus?

    A. millimeter

    B. centimeter

    C. meter

    D. kilometer

9. Which unit is best to measure the length of your tooth?

    A. millimeter

    B. centimeter

    C. meter

    D. kilometer

10. Use a centimeter ruler to measure the length, to the nearest centimeter, of the toy baseball bat below.

Answer _____

11. Juan measured the school football field and said it is about 100 units long. What metric unit of length did he use to measure the football field?

Answer _____

EXTENDED-RESPONSE QUESTION

12. Polly was meeting her friend, Indira, for lunch. She called Indira on her cell phone and said she was 500 kilometers away and would be at the restaurant in a few minutes.

Part A   Did Polly use the right metric unit? Why or why not?

_____

_____

Part B   What metric unit did Polly most likely mean?

_____

33 Metric Units of Mass

4.M.4, 4.M.5

Getting the Idea

When you want to know the amount of matter in an object, you measure its **mass**.

Metric Units of Mass
1,000 grams (g) = 1 kilogram (kg)

A paper clip has a mass of about 1 **gram**.

A pair of shoes has a mass of about 1 **kilogram**.

To measure the mass of an object, you can use a balance scale.

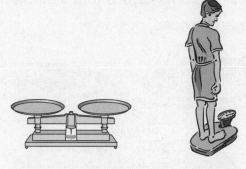

EXAMPLE 1

What is the mass of a pencil?

STRATEGY   Use a balance scale. Experiment with gram and kilogram masses until the trays are even.

STEP 1   Put a pencil on one side of a balance scale.

Put a 1-kilogram mass on the other side.

Check if the trays are even.

The trays are uneven. The pencil does not have a mass of 1 kilogram.

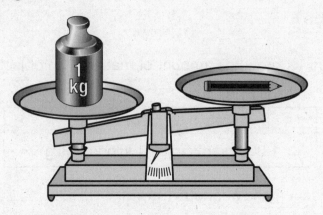

STEP 2   Try 1-gram masses.

Use 1-gram masses until the trays are even.

It takes four 1-gram masses to make the trays even.

The scale is balanced.

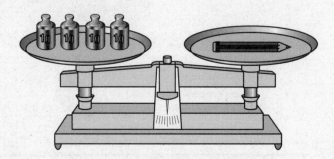

SOLUTION   **The mass of a pencil is about 4 grams.**

EXAMPLE 2

Erica is reading a paperback book. Which is the best estimate for the mass of a paperback book?

500 kilograms   500 grams

STRATEGY   **Think about the metric units of mass.**

STEP 1   Can the book have a mass of 500 kilograms?

1 kilogram is about the mass of a pair of shoes.

500 kilograms is the mass of about 500 pairs of shoes.

That is too much.

STEP 2   Can the book have a mass of 500 grams?

1 paper clip has a mass of 1 gram.

500 grams for the book could be right.

SOLUTION   **The mass of a paperback book is about 500 grams.**

COACHED EXAMPLE

The mass of 1 nickel is about 5 grams. What is the mass of 40 nickels, in grams?

THINKING IT THROUGH

The mass of 1 nickel is about _____ grams.

Multiply _____ nickels by _____ grams.

Compute.

_____ × _____ = _____

The mass of 40 nickels is about _____ grams.

Lesson Practice

Choose the correct answer.

1. What is the best estimate for the mass of a cat?

   A. 5 grams

   B. 50 grams

   C. 5 kilograms

   D. 50 kilograms

2. Which object would best be measured in grams?

   A. a toy car

   B. a television

   C. a bed

   D. an elephant

3. Mrs. Jonas used a scale to find the mass of some bananas.

   Which is closest to the mass of the bananas?

   A. 2 grams

   B. 20 grams

   C. 2 kilograms

   D. 20 kilograms

4. Which has a mass of about 200 kilograms?

A.

B.

C.

D.

5. Which is the closest estimate for the mass of a bowling ball?

   A. 1 kilogram

   B. 5 kilograms

   C. 10 grams

   D. 500 grams

6. Which has a mass of about 50 grams?

   A. a marker

   B. a pair of boots

   C. an apple

   D. a lunchbox

7. Which tool is best for measuring the mass of an apple?

   A. tape measure

   B. balance scale

   C. thermometer

   D. measuring cup

8. What metric unit would be best to measure the mass of a horse?

   Answer _____

9. Ralph measured the mass of his soccer ball. He said that its mass is about 400 units. What metric unit of mass did Ralph most likely use?

   Answer _____

34 Metric Units of Capacity

4.M.6, 4.M.7

Getting the Idea

When you want to know how much liquid a container can hold, you measure its liquid volume, or **capacity**.

You can measure capacity using milliliters or liters.

1 **milliliter** is a few drops of water.

1 **liter** is about four cups.

Some tools used to measure capacity are a graduated cylinder, a beaker, and a dropper.

When determining the best unit for measuring capacity, use the unit that makes the most sense.

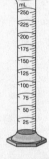

Graduated cylinder

Beaker

Dropper

EXAMPLE 1

The beaker has water in it.

How much water is in the beaker?

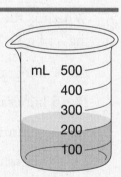

STRATEGY   **Find the amount of water in the beaker.**

    STEP 1   Look at the marks on the beaker.

                Each mark is another 100 milliliters.

    STEP 2   Read the mark that the water comes up to.

                The water stops at the 200 milliliters mark.

SOLUTION   **There are 200 milliliters in the beaker.**

EXAMPLE 2

Which unit of measure is best for measuring the amount of milk in a jug?

milliliter   liter

STRATEGY   **Think about the metric units of capacity.**

A milliliter is a few drops of water. It is too small.
A liter is about 4 cups.

SOLUTION   **The best unit of measure for measuring the amount of milk in a jug is a liter.**

COACHED EXAMPLE

The beaker has liquid in it.

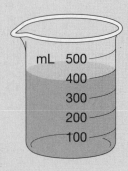

How much liquid is in the beaker?

THINKING IT THROUGH

Read the marks on the beaker.

Each mark is 100 _____.

The liquid stops at the _____ milliliters mark.

There are _____ of liquid in the beaker.

Lesson Practice

Choose the correct answer.

1. Katia poured some olive oil in a beaker.

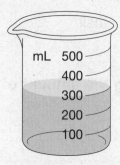

   How much olive oil is in the beaker?

   A. 100 milliliters

   B. 200 milliliters

   C. 300 milliliters

   D. 400 milliliters

2. Which unit of measure is best for measuring the amount of tea in a teacup?

   A. liter

   B. gram

   C. meter

   D. milliliter

3. Which is the measure of how much liquid a container holds?

   A. length

   B. mass

   C. weight

   D. capacity

4. Which unit of measure is best for measuring the amount of water in a sink?

   A. gram

   B. liter

   C. meter

   D. milliliter

5. Which tool is best for measuring a few drops of food coloring for a recipe?

   A. dropper

   B. thermometer

   C. balance scale

   D. ruler

6. Marta used some hot sauce on her tacos. Which unit of measure is best for measuring the amount of hot sauce she used?

   Answer _____

7. Eli bought a large bottle of apple juice. Which unit of measure is best for measuring the amount of apple juice he bought?

   Answer _____

35 Make Change

4.M.8

Getting the Idea

Money is used to buy things. Below are the coins and bills most often used in the United States.

Penny	Nickel	Dime	Quarter	Half Dollar
1¢	5¢	10¢	25¢	50¢
$0.01	$0.05	$0.10	$0.25	$0.50

Dollar	Five Dollars	Ten Dollars	Twenty Dollars
$1.00	$5.00	$10.00	$20.00

When you buy something at a store, you do not usually pay with the exact amount of money. You often give the cashier more money than the cost of your purchase and then get back change.

One way to make change is to count up from the price of the item to the amount of money you gave to the cashier.

EXAMPLE 1

DeShawn bought a book for $8.59. He gave the cashier a $10 bill. How much change should DeShawn receive from the cashier?

STRATEGY   **Count up from the price to the amount given to the cashier.**

STEP 1   Start with the price of the book and count up.

The book costs $8.59 and DeShawn gave the cashier $10.00.

$8.59, $8.60, $8.65, $8.75, $9.00, $10.00

STEP 2   Add up the values of the bill and coins.

$1.00, $1.25, $1.35, $1.40, $1.41

SOLUTION   **DeShawn should receive $1.41 in change from the cashier.**

You can also make change by subtracting the amount of the item from the total amount given to the cashier. Be sure to include the decimal point and dollar sign in the difference.

EXAMPLE 2

Lana bought a sweater for $27.36. She gave the cashier $30.00. How much change should Lana receive?

STRATEGY   **Subtract as you would with whole numbers.**

Line up the decimal points. Subtract to find the change.

Regroup if necessary.

$$\begin{array}{r} \overset{9\ \ 9}{\underset{}{2\ \cancel{10}\ \cancel{10}\ 10}} \\ \$\cancel{3}\cancel{0}.\cancel{0}\cancel{0} \\ -\ 2\ 7\ .\ 3\ 6 \\ \hline \$\ \ 2\ .\ 6\ 4 \end{array}$$

SOLUTION   **Lana should receive $2.64 in change.**

COACHED EXAMPLE

Malik bought a pen for $4.29. He paid with a $5 bill. How much change should Malik receive from the cashier?

THINKING IT THROUGH

Start with the price of the pen and count up using coins and bills from the least to the greatest.

$4.29, $4.30, $_____, $_____, $_____, $_____

What coins are needed to count up to $5 from $4.29?

_____ penny, _____ dime(s), _____ quarter(s)

Add up the values of the coins.

_____ + _____ + _____ + _____ + _____ = _____

Malik will receive _____ in change from the cashier.

Lesson Practice

Choose the correct answer.

1. Lori bought a magazine for $2.79. She paid with a $5 bill. Which shows the correct change that Lori should receive from the cashier?

   A.

   B.

   C.

   D.

2. Rico's dinner cost $6.35. He paid with a $10 bill. How much change should Rico receive?

   A. $3.65

   B. $3.75

   C. $4.65

   D. $4.75

3. Tracy buys lunch for her family for $15.75. If she pays with a $20 bill, how much will Tracy receive in change?

   A. $3.75

   B. $4.25

   C. $5.25

   D. $6.25

4. Heidi bought a pack of stickers for $0.63. She paid with a $1 bill. If she received the correct change, which coins could Heidi have received?

A.

B.

C.

D.

5. Jalen bought a bottle of juice for $1.45. He gave the cashier $2.00 in bills. How much should Jalen receive in change?

A. $3.45

B. $2.45

C. $1.55

D. $0.55

6. Joanne used a $5 bill to buy a bunch of bananas. The bananas cost $3.45. How much change should Joanne receive?

A. $1.45

B. $1.55

C. $2.45

D. $2.55

7. Aaron bought some apples and some pears at a local farm stand. The apples cost $4.75 and the pears cost $4.50. Aaron paid with a $10 bill. How much change should Aaron receive?

Answer _____

8. Terri's lunch cost the same amount of money as two $1 bills, four quarters, and two dimes. Terri paid with a $5 bill. How much money should Terri receive in change?

Answer _____

EXTENDED-RESPONSE QUESTION

9. Erin needs to buy some school supplies. She has $4.00 to spend on supplies. Prices for supplies are shown on the price list below.

School Supplies Price List	
pen	$3.75
pencil	$1.30
eraser	$0.40
sticky notes	$2.20

Part A What three different items can Erin buy that together total less than $4.00?

_____

Part B If Erin buys these three items, what change will she receive?

_____

_____

_____

36 Elapsed Time

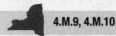

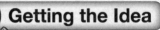

Elapsed time is the amount of time that has passed from a beginning time to an end time. The following units of time can help you solve problems with elapsed time.

Units of Time
1 **minute** = 60 **seconds**
1 **hour** = 60 minutes
1 **day** = 24 hours
1 **week** = 7 days
1 **year** = 12 **months**

EXAMPLE 1

Shawna began her homework at 4:30 P.M. She finished at 6:00 P.M. How long did Shawna do homework?

STRATEGY   Find the elapsed time.

    STEP 1   How many hours is it from 4:30 to 5:30?

          There is 1 hour from 4:30 to 5:30.

    STEP 2   How many minutes is it from 5:30 to 6:00?

          There are 30 minutes, or $\frac{1}{2}$ hour, from 5:30 to 6:00.

SOLUTION   **Shawna did homework for 1 hour 30 minutes.**

EXAMPLE 2

Soccer practice began at 8:30 A.M. It lasted 2 hours. What time did soccer practice end?

STRATEGY   **Count on 2 hours from 8:30 A.M.**

From 8:30 A.M. to 9:30 A.M. is 1 hour.

From 9:30 A.M. to 10:30 A.M. is 2 hours.

SOLUTION   **Soccer practice ended at 10:30 A.M.**

You can use a **calendar** to find elapsed time when days, weeks, and months are involved.

EXAMPLE 3

Meg left for vacation on April 16. She will be gone for 1 week. On what date will Meg return?

April						
Sunday	Monday	Tuesday	Wednesday	Thursday	Friday	Saturday
				1	2	3
4	5	6	7	8	9	10
11	12	13	14	15	**16**	17
18	19	20	21	22	23	24
25	26	27	28	29	30	

STRATEGY   **Use the calendar.**

STEP 1   What day did Meg leave?

Meg left on April 16. Mark this date on the calendar.

STEP 2   How many days are in 1 week?

There are 7 days in 1 week.

STEP 3   Count on 7 days.

From April 16 to April 17 is 1 day.

From April 17 to April 18 another day.

Seven days, or 1 week, after April 16 is April 23.

SOLUTION   **Meg will return on April 23.**

COACHED EXAMPLE

Christine's aunt arrived at her house on May 11. She left Christine's house 9 days later. On what date did Christine's aunt leave?

May						
Sunday	Monday	Tuesday	Wednesday	Thursday	Friday	Saturday
1	2	3	4	5	6	7
8	9	10	11	12	13	14
15	16	17	18	19	20	21
22	23	24	25	26	27	28
29	30	31				

THINKING IT THROUGH

Mark the date on the calendar when Christine's aunt arrived.

She arrived on Wednesday, May _____.

From May 11 to May 12 is 1 _____.

Count 8 more _____ on the calendar.

Christine's aunt left on May _____.

Lesson Practice

Choose the correct answer.

1. Ehud left home at 1:00 P.M. and returned at 4:00 P.M. How long was Ehud away from home?

   A. 4 hours 30 minutes

   B. 4 hours

   C. 3 hours 30 minutes

   D. 3 hours

2. Alana started studying at 3:30 P.M. and finished at 6:00 P.M. How long did Alana study?

   A. 2 hours 30 minutes

   B. 3 hours

   C. 3 hours 30 minutes

   D. 4 hours

3. At 7:00 P.M., Willa finished watching a movie that lasted 1 hour 30 minutes. At what time did the movie start?

   A. 5:00 P.M.

   B. 5:30 P.M.

   C. 6:00 P.M.

   D. 6:30 P.M.

4. A play started at 2:30 P.M. and ended at 5:00 P.M. How long was the play?

   A. 60 minutes

   B. 1 hour 30 minutes

   C. 2 hours

   D. 2 hours 30 minutes

Use the calendar for questions 5 and 6.

November						
Sun	Mon	Tues	Wed	Thurs	Fri	Sat
	1	2	3	4	5	6
7	8	9	10	11	12	13
14	15	16	17	18	19	20
21	22	23	24	25	26	27
28	29	30				

5. Jerome went to New York City the morning of November 9. He returned in the morning 5 days later. On what date did Jerome return?

   A. Nov. 4   **C.** Nov. 14

   B. Nov. 5   **D.** Nov. 16

6. Harry's birthday is on November 11. Exactly 2 weeks later is Thanksgiving. On what date is Thanksgiving?

   A. Nov. 18   **C.** Nov. 24

   B. Nov. 19   **D.** Nov. 25

Use the calendar and information below for questions 7 and 8.

On July 10, Sophie's airplane left the airport at 3:00 P.M. The airplane landed at 6:00 P.M. the same day.

July						
Sun	Mon	Tues	Wed	Thurs	Fri	Sat
			1	2	3	4
5	6	7	8	9	10	11
12	13	14	15	16	17	18
19	20	21	22	23	24	25
26	27	28	29	30	31	

7. How long was Sophie's airplane flight?

   Answer _____

8. Sophie returned home at 3:00 P.M. on July 16. How many days was Sophie gone?

   Answer _____

EXTENDED-RESPONSE QUESTION

9. Roberta wakes up at 7:00 A.M. She brushes her teeth, gets dressed, and eats breakfast before going to school. Roberta arrives at school at 9:00 A.M.

   Part A How much time passes from when Roberta wakes up to when she arrives at school?

   _____

   Part B Roberta goes to her mathematics class when she arrives at school. The class lasts one hour. At what time does the class end?

   _____

4 Review

1 Donna wants to measure the length of a pencil. Which unit of measure is **best** for Donna to use?

A inch   C foot

B yard   D mile

2 Ashley poured some milk into a sippy cup for her little sister. Which unit of measure is **best** for measuring the amount of milk in a sippy cup?

A gram   C meter

B milliliter   D liter

3 Lenny had $5.00 in his wallet. He bought a magazine for $3.65. How much change did Lenny receive when he paid for the magazine?

A $2.45   C $1.45

B $2.35   D $1.35

4 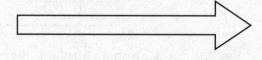 Use your ruler to help you solve this problem.

What is the length, in inches, of this arrow?

A 2 inches   C $2\frac{1}{2}$ inches

B $2\frac{1}{4}$ inches   D $2\frac{3}{4}$ inches

5 Mecia left for vacation with her family on May 9. They returned May 20.

MAY						
Sun	Mon	Tue	Wed	Thu	Fri	Sat
						1
2	3	4	5	6	7	8
9	10	11	12	13	14	15
16	17	18	19	20	21	22
23	24	25	26	27	28	29
30	31					

How long was their vacation?

A 1 week   C 9 days

B 2 weeks   D 11 days

6 What is the **best** tool to use to measure the mass of this book?

A balance scale

B ruler

C measuring cup

D tape measure

7 Max is 4 feet 10 inches tall. How tall is he in inches?

A 58 inches

B 50 inches

C 46 inches

D 42 inches

8 Use your ruler to help you solve this problem.

What is the length of this paper clip to the nearest centimeter?

A 5 centimeters

B 4 centimeters

C 3 centimeters

D 2 centimeters

9 Ms. Evans left New York City for a business meeting in Philadelphia at 6:30 A.M. She arrived at the meeting at 9:00 A.M. How long did it take Ms. Evans to reach the meeting?

A 2 hours

B 2 hours 30 minutes

C 3 hours

D 3 hours 30 minutes

10 Which unit of measure is **best** for measuring the amount of juice in a juice box?

A liter

B gram

C milliliter

D millimeter

11 Tai left for vacation on July 12. She will be gone for 2 weeks.

JULY						
Sun	Mon	Tue	Wed	Thu	Fri	Sat
				1	2	3
4	5	6	7	8	9	10
11	12	13	14	15	16	17
18	19	20	21	22	23	24
25	26	27	28	29	30	31

On what date will Tai return?

A July 14

B July 22

C July 24

D July 26

12 Laura bought a drink for $1.30. She gave the cashier $2.00. Which shows the money Laura could receive in change?

A

B

C

D

---

13 What is the **best** estimate for the mass of a cell phone?

A   10 grams

B   100 grams

C   500 grams

D   1,000 grams

14 What is the **best** tool to use to measure the length of a room?

A   ruler

B   balance scale

C   tape measure

D   thermometer

15 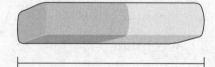 Use your ruler to help you solve this problem.

What is the length, in inches, of this eraser?

A $1\frac{1}{2}$ inches

B   2 inches

C $2\frac{1}{2}$ inches

D   3 inches

16 Casey arrived at the mall at 3:30 P.M. She left the mall at 8:00 P.M. How long was Casey at the mall?

Answer _____

17 Mr. Harris is 6 feet 4 inches tall. How tall is Mr. Harris in **inches**?

Show your work.

Answer _____ inches

18 Josie bought a newspaper for $1.50 and groceries for $7.89. She gave $10.00 to the cashier.

Part A

How much money did Josie receive in change?

Answer $ _____

Part B

What coins would Josie receive in change if she received the **fewest** number of coins possible?

Answer _____

5 Statistics and Probability

			NYS Math Indicators
Lesson 37	Formulate Questions and Use Surveys.	242	3.S.1*, 3.S.2*
Lesson 38	Display Data	247	4.S.3
Lesson 39	Collect and Record Data	253	4.S.1**, 4.S.2**
Lesson 40	Analyze Data and Make Predictions.	259	4.S.5, 4.S.6
Lesson 41	Line Graphs.	266	4.S.4
	Strand 5 Review	271	

* Grade 3 May–June Indicators   ** Grade 4 May–June Indicators

37 Formulate Questions and Use Surveys

 3.S.1, 3.S.2

Getting the Idea

A **survey** is a way of collecting **data** by asking people questions. You can use a table to collect the data and use a graph to display it. You can use a **tally** to record data, where each | represents 1 and ⅢⅡ represents 5.

EXAMPLE 1

The table shows the results of Hanna's survey.

Which question did Hanna most likely ask?

A. Do you like to watch sports?

B. What is your favorite type of movie?

C. What time does the game start?

D. Which is your favorite sport to watch?

Hanna's Survey

Sport	Tally
Baseball	ⅢⅡ ⅢⅡ
Football	ⅢⅡ ⅢⅡ |||
Basketball	ⅢⅡ ⅢⅡ ⅢⅡ |||
Soccer	ⅢⅡ ||

STRATEGY Review each choice.

STEP 1 Look at Choice A.

This question would have the answers of yes and no.

STEP 2 Look at Choice B.

The data in the table is not about movies.

STEP 3 Look at Choice C.

The data in the table is not about time.

STEP 4 Look at Choice D.

The data in the table is about types of sports and the number of students that like them.

This question makes sense.

SOLUTION Hanna most likely asked the question in Choice D, "Which is your favorite sport to watch?"

You can create a **pictograph** to display data. A pictograph uses symbols to represent numbers. The **key** tells how many each symbol represents. When you draw a pictograph, use a key that makes the graph easy to create and read.

EXAMPLE 2

Carla surveyed the number of books that four classes read for a book-a-thon.

Books Read

Ms. Dylan	Mr. Mitchell	Mrs. Lewis	Ms. Seger
26	32	28	20

Make a pictograph to represent the data in the table.

STRATEGY   Pick a key that makes the graph easy to create and read.

STEP 1   Look at the data.

Since all of the numbers are divisible by 4, use 4 for the key.

STEP 2   Pick a symbol and a title.

Use a book for the symbol. Title the graph "Books Read."

STEP 3   Divide each number of books by 4 to find how many symbols to use.

$26 \div 4 = 6$ R2, so use $6\frac{1}{2}$ symbols.

$32 \div 4 = 8$ $28 \div 4 = 7$ $20 \div 4 = 5$

STEP 4   Make the pictograph.

Make sure that you add a title and make a key.

Books Read

Ms. Dylan	📖📖📖📖📖📖📖
Mr. Mitchell	📖📖📖📖📖📖📖📖
Mrs. Lewis	📖📖📖📖📖📖📖
Ms. Seger	📖📖📖📖📖

Key: Each 📖 = 4 books

SOLUTION   **The pictograph is shown in Step 4.**

COACHED EXAMPLE

Scarlett did a survey. She made a pictograph to show the data.

Favorite Ice Cream Flavors

Vanilla	🍦 🍦 🍦 🍦 🍦 🍦 🍦
Chocolate	🍦 🍦 🍦 🍦 🍦 🍦
Strawberry	🍦 🍦 🍦 🍦
Mint	🍦 🍦 🍦

Key: Each 🍦 = 4 votes

Make a table to show the same data as in the pictograph.
Write a question that Scarlett most likely asked for the survey.

THINKING IT THROUGH

Make a table. Write the title and names of flavors.

Title: _____

Flavor	Number of Votes

Find the number of votes for each flavor.

What does each symbol represent? _____ votes

How many students voted for vanilla? _____ × 4 = _____

How many students voted for chocolate? _____ × 4 = _____

How many students voted for strawberry? _____ × 4 = _____

How many students voted for mint? _____ × 4 = _____

Write the number of votes in the table for each flavor.

Write a question that Scarlett most likely asked.

_____

Lesson Practice

Choose the correct answer.

Use the table for questions 1–3.

Jin did a survey and made the table below.

Favorite Lunch Food

Food	Number of Students
Hamburger	12
Pizza	20
Hot Dogs	12
Grilled Cheese	8

1. Which question did Jin most likely ask?

   A. How much did you spend on lunch?

   B. Do you like to have pizza for lunch?

   C. What is your favorite lunch food?

   D. At what time do you eat lunch?

2. How many students did Jin survey?

   A. 20

   B. 32

   C. 50

   D. 52

3. Which pictograph shows the same data as the table?

   Favorite Lunch Food

   A.

Hamburger	🍪🍪🍪🍪🍪
Pizza	🍪🍪🍪🍪🍪🍪🍪🍪🍪🍪
Hot Dogs	🍪🍪🍪🍪🍪
Grilled Cheese	🍪🍪🍪🍪

   Key: Each 🍪 = 4 students

   Favorite Lunch Food

   B.

Hamburger	🍪🍪
Pizza	🍪🍪🍪
Hot Dogs	🍪🍪🍪🍪🍪
Grilled Cheese	🍪🍪🍪

   Key: Each 🍪 = 4 students

   Favorite Lunch Food

   C.

Hamburger	🍪🍪🍪
Pizza	🍪🍪🍪🍪🍪
Hot Dogs	🍪🍪🍪
Grilled Cheese	🍪🍪

   Key: Each 🍪 = 4 students

   Favorite Lunch Food

   D.

Hamburger	🍪🍪🍪
Pizza	🍪🍪🍪
Hot Dogs	🍪🍪
Grilled Cheese	🍪🍪🍪🍪🍪

   Key: Each 🍪 = 4 students

EXTENDED-RESPONSE QUESTION

4. Mrs. Thomas surveyed some students and made a graph.

Favorite Type of Bread

White	🍞 🍞 🍞 🍞 🍞
Wheat	🍞 🍞 🍞 🍞
Rye	🍞 🍞

Key: Each 🍞 = 2 students

Part A Write a question that Mrs. Thomas most likely asked.

_____

Part B Complete the table below to show the results of Mrs. Thomas's survey.

Title _____

Type of Bread	Number of Students

38 Display Data

4.S.3

Getting the Idea

You can create a table to display data.

EXAMPLE 1

Sarah surveyed her classmates. Seven students chose the Knicks, 9 students chose the Yankees, 3 students chose the Rangers, and 4 students chose the Giants. Make a table to record her data.

STRATEGY Record the data in a table.

STEP 1   Name the table.

Sarah probably asked her classmates to name their favorite sports team so the title should be "Favorite New York Sports Team."

STEP 2   Write the four sports teams in the first column.

Record the number for each team.

Favorite New York Sports Team

Team	Number
Knicks	7
Yankees	9
Rangers	3
Giants	4

SOLUTION   The table of Sarah's data is shown in Step 2.

You can create a **bar graph** to display data. A bar graph is used to compare data by using bars of different lengths. To draw a bar graph, choose an **interval** on the vertical axis that makes the graph easy to read. The distance between intervals should be equal.

EXAMPLE 2

The table shows the number of students that received each grade from A to F on a math test.

Grades on a Math Test

Grade	A	B	C	D	F
Number of Students	18	24	30	12	8

Make a bar graph to represent the data.

STRATEGY   Pick an interval. Then plot the data.

STEP 1   Pick an interval that will make the graph easy to read.

All the numbers are multiples of 2. An interval of 2 will make the graph too large. 4 will make the graph easy to read.

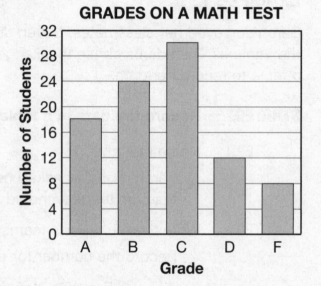

STEP 2   Make the outline of the graph and label the axis. Give the graph a title.

Label the horizontal axis "Grade." Put the letters A through F on the axis.

Label the vertical axis "Number of Students."

Title the graph "Grades on a Math Test."

STEP 3   Make the bar graph.

Draw bars for each grade.

The bars for A and C end halfway between intervals.

SOLUTION   The bar graph is shown in Step 3.

COACHED EXAMPLE

The table shows the number of used cars sold by each salesperson at a used car lot.

Cars Sold Last Month

Salesperson	Julio	Gary	Rina	Adam
Cars Sold	25	15	30	20

Make a bar graph to represent the data.

THINKING IT THROUGH

Write a title for your graph. _____

What should you label the horizontal axis? _____

What should you label the vertical axis? _____

Choose an interval to use. Since all of the numbers are divisible by 5, I chose intervals of _____.

Make your bar graph.

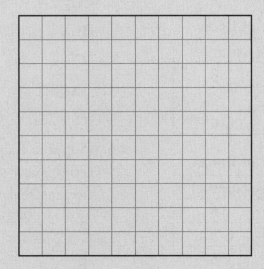

Lesson Practice

Choose the correct answer.

1. Tana surveyed her classmates about their favorite type of book. 3 students chose adventure, 5 students chose mystery, and 10 students chose science fiction. Which table correctly displays Tana's data?

   Favorite Type of Book

   A.
Book Type	Number
Adventure	10
Mystery	5
Science Fiction	3

   Favorite Type of Book

   B.
Book Type	Number
Adventure	5
Mystery	3
Science Fiction	10

   Favorite Type of Book

   C.
Book Type	Number
Adventure	3
Mystery	5
Science Fiction	10

   Favorite Type of Book

   D.
Book Type	Number
Adventure	10
Mystery	3
Science Fiction	5

Use the pictograph for questions 2–4.

Cupcakes Sold

Timmy	🧁🧁🧁🧁🧁🧁
Amy	🧁🧁🧁🧁🧁
Melissa	🧁🧁🧁🧁🧁🧁
Scott	🧁🧁🧁🧁🧁

Key: Each 🧁 = 4 cupcakes sold

2. Who sold exactly 24 cupcakes?

   A. Timmy

   B. Amy

   C. Melissa

   D. Scott

3. How many cupcakes does each symbol represent?

   A. 2

   B. 4

   C. 6

   D. 8

4. Suppose Ralph sold 18 cupcakes. How many symbols would you need to represent Ralph's data?

   A. 4

   B. $4\frac{1}{2}$

   C. 5

   D. $5\frac{1}{2}$

Use the bar graph for questions 5 and 6.

The graph shows the numbers of seashells found by four students while on vacation at the beach.

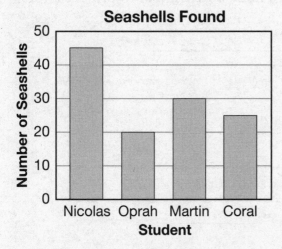

Seashells Found

5. Which student found 20 seashells?

   A. Nicholas

   B. Oprah

   C. Martin

   D. Coral

6. How many seashells did Coral find?

   A. 45

   B. 30

   C. 25

   D. 20

7. Nolan conducted a survey in his fourth-grade class. He recorded the data in the table below.

Fruit	Number
Apple	6
Orange	3
Banana	4
Pineapple	7

What survey question did Nolan most likely ask?

Answer _____

8. Tania made a bar graph to show the pets her classmates have. 5 students have cats, 7 students have dogs, 4 students have fish, and 2 students have turtles. Which pet will have the longest bar on Tania's graph?

Answer _____

EXTENDED-RESPONSE QUESTION

9. Rupert asked his classmates which activity they liked best. 2 students chose biking, 4 students chose hiking, 8 students chose camping, and 6 students chose swimming.

Part A Make a pictograph to show Rupert's results. Remember to provide a title and key for your graph.

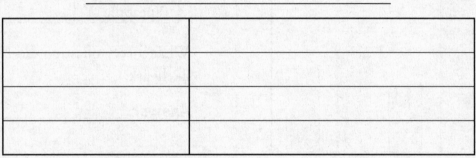

Key: Each _____ = _____

Part B Make a bar graph to show Rupert's results. Remember to provide a title, label both axes, and choose an interval that works for the data.

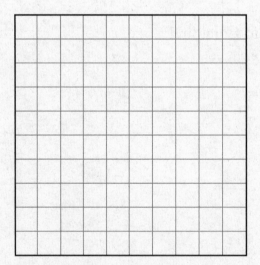

39 Collect and Record Data

 4.S.1, 4.S.2

Getting the Idea

Remember, a survey is a way to collect data. It is conducted by asking people questions and recording their answers. Once the data is collected, it can be organized in a table or graph.

EXAMPLE 1

Lenny surveyed his fourth grade class about the way they get to school each day. The results of his survey are shown below.

bus, bus, walk, car, car, bus, car, walk, bus, car, bus, bus, car, car, bus, walk, car, bus, bus, bus, walk, car, walk, bus, bus, bus, car, bus

Lenny wants to find which way is used the most. What did Lenny learn from the results of his survey?

STRATEGY   Organize the data into a table. Then analyze the data.

STEP 1   Make a table to organize the data.

There were three answers choices: bus, walk, car.

Count the number of votes for each choice.

Record the numbers in your table.

Ways Fourth Graders Get to School

Way	Number of Fourth Graders
Bus	14
Car	9
Walk	5

STEP 2   Look at the data and analyze the results.

The greatest number of fourth graders ride the bus.

SOLUTION   **Lenny learned that most fourth graders get to school each day on the bus.**

You can also perform an experiment, collect data, and organize your results in a table or bar graph.

EXAMPLE 2

Rachael spun a four-color spinner 30 times. The results of her experiment are shown below.

red, red, yellow, red, blue, blue, green, red, yellow, green, red, blue, blue, green, yellow, red, yellow, yellow, red, red, yellow, red, yellow, green, red, red, yellow, green, blue, blue

Rachael guessed that the spinner would land on red most often. Did Rachel's results support her guess?

STRATEGY   **Organize the data in a bar graph. Then analyze the data.**

STEP 1   Make a bar graph showing the data.

Title and label the graph.

Choose a scale that works for the data.

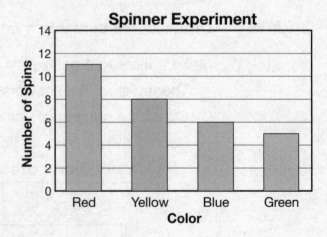

STEP 2   Look at the height of the bars in the bar graph.

The bar for red is the tallest bar.

STEP 3   Compare the results with Rachael's guess.

The spinner landed on red more times than any other color.

Rachael's guess was supported by this experiment.

SOLUTION   **Rachael's results did support her guess.**

EXAMPLE 3

On Friday, Mr. Regis drew the following table on the board to show last week's absences.

Last Week's Absences

Day	Number of Students
Monday	3
Tuesday	5
Wednesday	4
Thursday	6
Friday	7

Mr. Regis asked, "How many students were absent last Thursday and Friday?"

STRATEGY   **Use addition to find the answer. Then make a plan to find the number of absent students this week.**

STEP 1   Find how many students were absent last Thursday.

There were 6 students absent last Thursday.

STEP 2   Find how many students were absent last Friday.

There were 7 students absent last Friday.

STEP 3   Add the two numbers.

6 + 7 = 13

There were 13 students absent last Thursday and Friday.

SOLUTION   **There were 13 students absent last Thursday and Friday.**

Remember, each | represents 1 tally and ||||| represents 5 tallies.

COACHED EXAMPLE

The following data shows the number of squirrels that four students observed on a nature walk. How many more squirrels did Eileen and Becky see than Thomas and Roy?

> Roy – ||||| |
> Thomas – ||||| ||||| ||||
> Eileen – ||||| |||||
> Becky – ||||| ||||| ||

THINKING IT THROUGH

Make a pictograph to show the data. Choose a title and a key.

_____

Key: Each _____ = _____

How many squirrels did Eileen see? _____

How many squirrels did Becky see? _____

Add Eileen and Becky's numbers: _____ + _____ = _____

How many squirrels did Thomas see? _____

How many squirrels did Roy see? _____

Add Thomas and Roy's numbers: _____ + _____ = _____

Subtract Thomas and Roy's number from Eileen and Becky's.

_____ – _____ = _____

Eileen and Becky saw _____ more squirrels than Thomas and Roy.

Lesson Practice

Choose the correct answer.

Use the table for questions 1 and 2.

The following table shows the results of a survey about favorite types of music taken at a local community center.

Favorite Music

Type of Music	Number of People
Country	18
Classical	9
Rock and Roll	24
Jazz	12
Rap	20

1. How many people's favorite music is country or jazz?

    A. 12   **C.** 20

    B. 18   **D.** 30

2. Which is a true statement about the results of this survey?

    A. Most people like jazz music the best.

    B. More people like rock and roll than classical music.

    C. Rap music is the least favorite music.

    D. The people surveyed like all types of music equally.

Use the pictograph for questions 3 and 4.

Zoo Animals

Bears	🐻 🐻 🐻 🐻 🐻 🐻
Elephants	🐻 🐻 🐻
Giraffes	🐻 🐻 🐻 🐻 🐻
Tigers	🐻 🐻 🐻 🐻

Key: Each 🐻 = 2 animals

3. How many more giraffes are there than tigers at the zoo?

    A. 1

    B. 2

    C. 3

    D. 4

4. Which conclusion **cannot** be made from this pictograph?

    A. There are more bears than giraffes at the zoo.

    B. Elephants are the most popular animals at the zoo.

    C. There are fewer tigers than bears at the zoo.

    D. There are 10 giraffes at the zoo.

Use the data for questions 5–7.

Favorite Granola Bars

chocolate chip, chocolate chip, peanut butter, oatmeal raisin,
peanut butter, chocolate chip, peanut butter, oatmeal raisin,
chocolate chip, peanut butter, oatmeal raisin, chocolate chip,
oatmeal raisin, chocolate chip, chocolate chip, peanut butter

5. Which table correctly shows the data?

A.

Favorite Granola Bars

Granola Bar	Number
Chocolate Chip	5
Peanut Butter	4
Oatmeal Raisin	7

C.

Favorite Granola Bars

Granola Bar	Number
Chocolate Chip	7
Peanut Butter	5
Oatmeal Raisin	4

B.

Favorite Granola Bars

Granola Bar	Number
Chocolate Chip	4
Peanut Butter	5
Oatmeal Raisin	7

D.

Favorite Granola Bars

Granola Bar	Number
Chocolate Chip	7
Peanut Butter	6
Oatmeal Raisin	3

6. How many people voted for their favorite granola bar in all?

Answer _____

7. What question could have been used for this survey?

Answer _____

40 Analyze Data and Make Predictions

4.S.5, 4.S.6

Getting the Idea

Tables and graphs help you organize data. Once your data is put into a table or graph, you can more easily compare and analyze the information.

EXAMPLE 1

At the fourth-grade-class fair, students won tokens for winning different games. The following is a pictograph showing how many tokens four students won. How many more tokens did Sammy win than Carlos?

Tokens Won

Sammy	◎◎◎◎◎◎◎
Carlos	◎◎◎◎◖
Manny	◎◎◎◎
Vladimir	◎◎◎◎◎

Key: Each ◎ = 10 tokens

STRATEGY Find the number of tokens that each person won. Then subtract.

STEP 1 Count the number of tokens for Sammy.

There are 7 tokens for Sammy. Each token represents 10 tokens.

$7 \times 10 = 70$

STEP 2 Count the number of tokens for Carlos.

There are 4 full tokens: $4 \times 10 = 40$.

There is 1 half token: $10 \div 2 = 5$.

$40 + 5 = 45$

STEP 3 Subtract Carlos's total from Sammy's total.

$70 - 45 = 25$

SOLUTION Sammy won 25 more tokens than Carlos.

You can use data in tables and graphs to make **predictions** about what might happen. A prediction is a guess based on information.

EXAMPLE 2

Ms. Jenkins asked 50 fourth-grade students to vote for their favorite flavor of ice cream. The graph below shows the results.

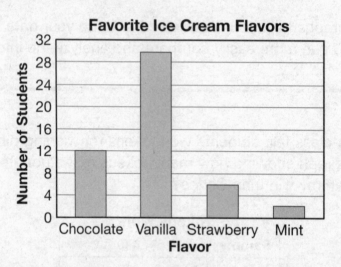

Predict which flavor would receive the most votes if Ms. Jenkins surveyed the remaining 50 fourth-grade students in the school.

STRATEGY   **Make a prediction based on the graph.**

STEP 1   Find which flavor received the most votes.
Vanilla received the most votes.

STEP 2   Make a prediction.
The bar graph shows that vanilla received the most votes when 50 fourth-grade students were surveyed. It will probably receive the most votes when the rest of the fourth-grade students are surveyed.

SOLUTION   **Vanilla will most likely receive the most votes when all the fourth-grade students are surveyed.**

COACHED EXAMPLE

The table shows the total number of baseball cards in four students' collections at the end of each week. Every week, the students add the same number of cards to their collections as the week before.

Total Number of Baseball Cards Collected at the End of Each Week

	Joe	Larry	Eva	Spencer
Week 1	20	22	14	28
Week 2	22	26	17	29
Week 3	24	30	20	30

If each student continues to collect the same number of baseball cards each week, who will have the least number of baseball cards by the end of Week 4?

THINKING IT THROUGH

How many cards does Joe add to his collection each week? _____

Add this number to Joe's total in Week 3.

24 + _____ = _____

How many cards does Larry add to his collection each week? _____

Add this number to Larry's total in Week 3.

30 + _____ = _____

How many cards does Eva add to her collection each week? _____

Add this number to Eva's total in Week 3.

20 + _____ = _____

How many cards does Spencer add to his collection each week? _____

Add this number to Spencer's total in Week 3.

30 + _____ = _____

Write the four totals for Week 4 in order from greatest to least.

_____ < _____ < _____ < _____

_____ will have the least number of baseball cards by the end of Week 4.

Lesson Practice

Choose the correct answer.

Use the pictograph for questions 1–4.

Museum Gift Shop Visitors

Monday	웃 웃 웃 웃 웃 〉
Tuesday	웃 웃 웃 웃 웃 웃 웃 웃
Wednesday	웃 웃 웃 웃 웃 웃 웃 〉
Thursday	웃 웃 웃
Friday	웃 웃 웃 웃 웃 웃 웃 웃

Key: Each 웃 = 10 visitors

1. How many people visited the gift shop on Tuesday and Thursday?

   A.   30

   B.   70

   C.   100

   D.   130

2. On which day did the most people visit the gift shop?

   A.   Monday

   B.   Tuesday

   C.   Wednesday

   D.   Friday

3. How many more people visited the gift shop on Monday than on Thursday?

   A.   25

   B.   20

   C.   15

   D.   10

4. How many more people visited the gift shop on Monday, Tuesday and Thursday than visited on Wednesday and Friday?

   A.   5

   B.   10

   C.   15

   D.   20

Use the bar graph for questions 5–8.

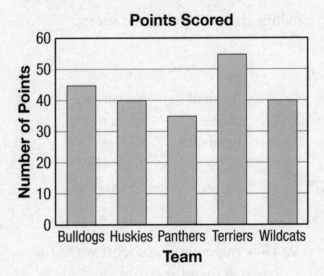

Points Scored

5. How many more points did the Bulldogs score than the Panthers?

   A. 5

   B. 10

   C. 15

   D. 20

6. Which sentence is true?

   A. The Panthers scored more points than the Wildcats.

   B. The Huskies scored fewer points than the Panthers.

   C. The Bulldogs scored more points than the Terriers.

   D. The Huskies and the Wildcats scored the same number of points.

7. Which team scored 5 points more than the Huskies?

   A. Bulldogs

   B. Terriers

   C. Panthers

   D. Wildcats

8. Which team scored the second greatest number of points?

   A. Terriers

   B. Huskies

   C. Bulldogs

   D. Panthers

Use the following data for questions 9–11.

Steven surveyed thirty fourth-grade students at random about their favorite sports. His results are shown below.

football	baseball	football	soccer	soccer
basketball	football	soccer	baseball	soccer
baseball	football	basketball	soccer	soccer
basketball	football	soccer	baseball	soccer
soccer	baseball	soccer	soccer	basketball
soccer	soccer	baseball	soccer	baseball

9. Which two sports were the least favorites, according to Steven's survey results?

   A. football and soccer

   B. basketball and baseball

   C. football and basketball

   D. baseball and soccer

10. How many more students voted for baseball than for football?

    A. 9

    B. 7

    C. 5

    D. 2

11. If Steven surveys the rest of the fourth-grade students, which sport do you predict will receive the most votes?

    A. football

    B. baseball

    C. basketball

    D. soccer

12. The table below shows the number of miles traveled by a car over a 5-hour period. If the ride is 6 hours, how many miles do you predict the car will travel?

Miles Traveled by a Car

Hours	1	2	3	4	5
Number of Miles	65	130	195	260	325

Answer _____

13. Thomas delivers newspapers every morning. Each week, he delivers 30 newspapers a day on five of the days, 50 newspapers a day on one of the days, and 70 newspapers a day on one of the days. Thomas says that Sundays are the busiest for him and Saturdays are the second busiest. How many newspapers does Thomas deliver each Saturday?

Answer _____

EXTENDED-RESPONSE QUESTION

14. The table below shows the amount in Carl's savings account at the end of each week.

Carl's Savings Account

Week	Total Amount of Money
1	$25
2	$50
3	$75
4	$100

Part A How much money does Carl put into his savings account each week?

_____

Part B If Carl continues to put the same amount of money into his saving account each week, how much money will Carl have in his account at the end of Week 6?

_____

Lesson

41 Line Graphs

 4.S.4

Getting the Idea

A **line graph** is used to compare data that change over time.

EXAMPLE 1

Tom recorded the outside temperature in the line graph below. How many degrees did the temperature increase from 8 A.M. to 11 A.M.?

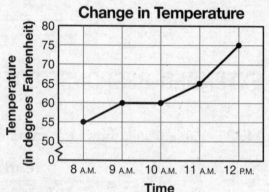

STRATEGY Read the temperatures at 8 A.M. and 11 A.M. Subtract to find the difference.

STEP 1 Find 8 A.M. on the axis labeled "Time."

Move up the line for 8 A.M. to the point.

Move left to the axis labeled "Temperature."

The temperature was 55°F at 8 A.M.

STEP 2 Find 11 A.M. on the axis labeled "Time."

Move up the line for 11 A.M. to the point.

Move left to the axis labeled "Temperature."

The temperature was 65°F at 11 A.M.

.STEP 3 Subtract.

65°F − 55°F = 10°F

SOLUTION The temperature increased 10°F from 8 A.M. to 11 A.M.

Sometimes the data in a line graph form a pattern. You can use a pattern to make a prediction.

EXAMPLE 2

The line graph below shows the number of miles traveled by a train over a 5-hour period. If the ride is 7 hours, how many miles do you predict the train will travel?

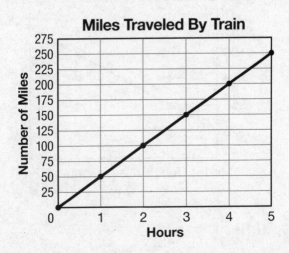

Miles Traveled By Train

STRATEGY   **Make a table. Find a pattern. Continue the pattern to make a prediction.**

STEP 1   Make a table to see if the graph shows a pattern.

Miles Traveled By a Train

Hours	0	1	2	3	4	5
Number of Miles	0	50	100	150	200	250

The number of miles increases by 50 miles each hour.

STEP 2   Use the data to predict.

If the train travels 2 more hours, it will have traveled 7 hours.

Skip count by 50 two times from 250.

250, 300, 350

SOLUTION   **If the ride is 7 hours, the train should travel 350 miles.**

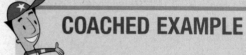

COACHED EXAMPLE

This line graph shows the amount of money in Carl's savings account each day.

Predict how much money Carl will have in his savings account on Saturday.

THINKING IT THROUGH

Make a table to show how much money Carl had each day.

Money in Carl's Account

Day	Amount (in dollars)
Monday	
Tuesday	
Wednesday	
Thursday	
Friday	

The pattern is _____ to each day's value.

What can you do to Friday's value to predict how much money Carl will have on Saturday? _____

Find the amount. _____ + _____ = _____

A good prediction is that Carl will have $_____ in his saving account on Saturday.

Lesson Practice

Choose the correct answer.

Use the graph for questions 1 and 2.

The following line graph shows the number of inches of snow that fell over 6 days.

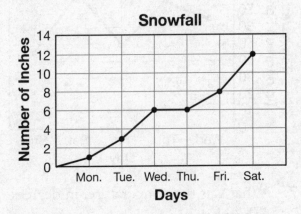

1. On what day did it snow 8 inches?

   A. Tuesday

   B. Wednesday

   C. Thursday

   D. Friday

2. Which two days had the same amount of snow?

   A. Friday and Saturday

   B. Wednesday and Thursday

   C. Tuesday and Wednesday

   D. Monday and Tuesday

Use the graph for questions 3 and 4.

The following line graph shows the amount of money in Carson's savings account each day one week.

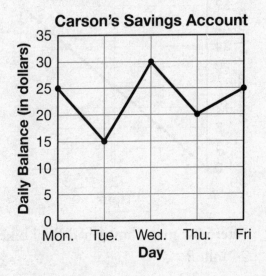

3. On which day was the daily balance the greatest?

   A. Monday

   B. Wednesday

   C. Thursday

   D. Friday

4. How much money was in the account on Tuesday?

   A. $15

   B. $20

   C. $25

   D. $30

Use the graph for questions 5 and 6.

The following line graph shows the total distance that Raul biked each hour.

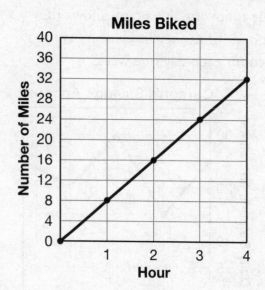

5. After how many hours had Raul biked 24 miles?

   A. 1 hour

   B. 2 hours

   C. 3 hours

   D. 4 hours

6. If the pattern in the line graph continues, how many miles will Raul travel after 6 hours?

   A. 16 miles

   B. 48 miles

   C. 54 miles

   D. 56 miles

Use the graph for questions 7 and 8.

The following line graph shows the daily high temperatures for a 5-day period.

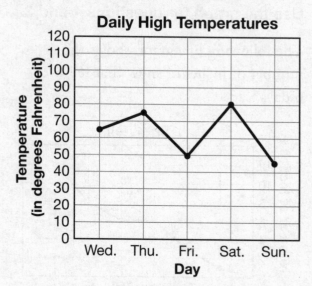

7. What was the temperature, in degrees Fahrenheit, on Friday?

   Answer _____

8. Which day had the lowest temperature?

   Answer _____

5 Review

1 Each Friday Mr. Pasqua divides his class into four teams to play math games. The number of games that four teams have won is shown in the bar graph below.

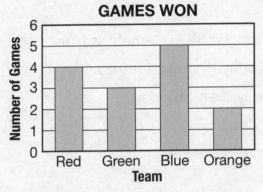

GAMES WON

Based on the information in the bar graph, which statement is true?

A The Blue team has won exactly half of the games.

B The Blue team has won more than twice as many games as the Green team.

C The Red and Green teams together have won exactly half of the games.

D The Blue and Orange teams have won more games than the Red and Green teams.

2 The table shows the number of push-ups Maya did each day this week.

MAYA'S PUSH-UPS

Day	Number of Push-Ups
Sunday	24
Monday	28
Tuesday	32
Wednesday	36

If the pattern continues, how many push-ups will Maya do Friday?

A 40

B 42

C 44

D 48

Use the pictograph for questions 3 and 4.

The pictograph shows the number of raffle tickets that were sold by Eduardo each day this week.

TICKETS SOLD

Monday	🎟️🎟️🎟️
Tuesday	🎟️🎟️🎟️🎟️🎟️
Wednesday	🎟️🎟️🎟️🎟️🎟️🎟️🎟️
Thursday	🎟️🎟️🎟️🎟️🎟️🎟️🎟️🎟️🎟️

KEY
🎟️ = 5 tickets sold

3 Based on the information in the pictograph, which statement is true?

A Eduardo sold 20 more tickets on Thursday than on Tuesday.

B Eduardo sold 4 more tickets Wednesday than Monday.

C Eduardo sold at least 20 tickets each day.

D Eduardo sold fewer than 100 tickets for the week.

4 If the pattern of tickets sold continues, how many tickets do you predict that Eduardo will sell Friday?

A 50

B 55

C 60

D 65

11 The table shows the number of pages that four students read yesterday.

PAGES READ

Student	Number of Pages Read
Pedro	20
Lisa	24
Tim	16
Natalie	32

Make a pictograph to show the number of pages that each student read.

Be sure to

- title the graph

- provide a key for the graph

- graph all the data

Title: _____

KEY
___ = _____

12 Jason, Tara, and Sam recorded the number of e-mails they received each day for three days. The table shows the number of e-mails each received.

E-MAILS RECEIVED

Student	Number of E-Mails		
	Friday	Saturday	Sunday
Jason	6	8	4
Tara	4	10	6
Sam	8	4	4

Part A

On the lines below, write the total number of e-mails each student received over the three days.

Jason _____ e-mails

Tara _____ e-mails

Sam _____ e-mails

Part B

On the grid, make a bar graph to show the total number of e-mails each student received.

Be sure to

- title the graph
- label both axes
- provide a scale for the graph
- graph all the data

13 The line graph shows the amount of money that Miguel earns at his part-time job for each hour he works.

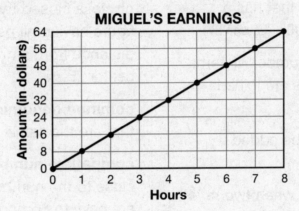

MIGUEL'S EARNINGS

Part A

How much money did Miguel earn after 5 hours?

Answer $ _____

Part B

Based on the data in the line graph, how much money will Miguel earn if he works 10 hours?

Answer $ _____

On the lines below, explain how you determined your answer.

_____

_____

_____

_____

Glossary

acute angle an angle that has a measure of less than 90° (Lesson 27)

add to find the total number of items when two more groups are joined (Lesson 4)

addend a number to be added (Lesson 4)

angle a figure formed when two rays meet at the same endpoint (Lesson 27)

area the number of square units needed to cover a region (Lesson 29)

array an arrangement of objects in rows and columns (Lesson 5)

associative property of multiplication property by which the grouping of factors does not change the product (Lesson 6)

bar graph a graph that shows data by using bars of different lengths (Lessons 38, 39)

base a bottom side or face of a geometric figure (Lesson 30)

calendar a chart that shows the days and weeks in a month or year (Lesson 36)

capacity the amount of liquid that a container can hold (Lesson 34)

center the point in a circle that is the same distance from all points (Lesson 28)

centimeter (cm) a metric unit of length; 100 centimeters = 1 meter (Lesson 32)

circle a closed two-dimensional figure having all points the same distance from a given point called the center (Lesson 28)

common denominators denominators that are the same (Lesson 15)

compatible numbers numbers that are close to the numbers in a problem and are easy to compute with (Lesson 21)

cone a solid figure with a circular base, a curved surface, and 1 vertex (Lesson 30)

cube a solid figure with 6 square faces, 12 edges, and 8 vertices (Lesson 30)

customary system the system of units of measure used in the United States (also called the English system) (Lesson 31)

cylinder a solid figure with 2 congruent circular faces and 1 curved surface (Lesson 30)

data information (Lesson 37)

day (d) a period of time; 1 day = 24 hours (Lesson 36)

decimal a number with a decimal point (Lesson 16)

decimal point (.) a period separating the ones from the tenths in a decimal (Lesson 16)

degree (°) a unit for measuring angles (Lesson 27)

denominator the number below the bar in a fraction; it tells the number of equal parts in all (Lesson 12)

difference the answer in a subtraction problem (Lesson 4)

divide to separate a number into equal groups (Lesson 9)

dividend the number to be divided (Lesson 9)

divisor the number by which the dividend is divided (Lesson 9)

edge a line segment where two faces of a solid figure meet (Lesson 30)

elapsed time the amount of time that passes from a beginning time to an end time (Lesson 36)

endpoints the points on the end of a line segment or ray (Lesson 26)

equation a statement that the values of two mathematical expressions are equal (Lesson 22)

equivalent fractions two or more fractions that have the same value (Lesson 13)

estimate to find an answer that is close to the exact answer (Lesson 21)

even number a whole number that has 0, 2, 4, 6, or 8, in the ones place (Lesson 6)

expanded form the sum of the values of all of the digits in a number (Lesson 1)

face a flat side of a three-dimensional figure (Lesson 30)

fact family a group of related facts, either addition and subtraction or multiplication and division, that use the same numbers (Lesson 9)

factors numbers that are multiplied to give a product (Lesson 5)

foot (ft) a customary unit of length; 1 foot = 12 inches (Lesson 31)

fraction a number that names part of a whole or group (Lesson 12)

gram (g) a metric unit of mass; 1,000 grams = 1 kilogram (Lesson 33)

hexagon a polygon with 6 sides (Lesson 28)

hour (h) a unit of time; 1 hour = 60 minutes (Lesson 36)

inch (in.) a customary unit of length; 12 inches = 1 foot (Lesson 31)

inequality a mathematical sentence that shows two values are not equal (Lesson 23)

input/output table a table that takes an input number and changes it to an output number using a rule (Lesson 25)

intersecting describes lines or line segments that meet or cross (Lesson 26)

interval the difference between numbers on an axis of a graph (Lesson 38)

is equal to (=) a symbol that shows that two quantities have the same value (Lessons 2, 17)

is greater than (>) a symbol that shows that the first quantity is greater than the second quantity (Lessons 2, 17)

is less than (<) a symbol that shows that the first quantity is less than the second quantity (Lessons 2, 17)

is not equal to a symbol that shows that two quantities do not have the same value (Lessons 2, 17)

key in a pictograph, it tells how many each symbol represents (Lesson 37)

kilogram (kg) a metric unit of mass; 1 kilogram = 1,000 grams (Lesson 33)

kilometer (km) a metric unit of length; 1 kilometer = 1,000 meters (Lesson 32)

length how long or tall something is (Lesson 31)

like fractions fractions with the same denominator (Lesson 25)

line goes in two directions without end, which is indicated by arrows on either end (Lesson 26)

line graph a graph that uses a line to show how something changes over time (Lesson 41)

line segment a part of a line with endpoints (Lesson 26)

liter (L) a metric unit of capacity; 1 liter = 1,000 milliliters (Lesson 34)

mass the amount of matter in an object (Lesson 33)

meter (m) a metric unit of length; 1 meter = 100 centimeters (Lesson 32)

metric system the system of units of measure most commonly used throughout the world (Lesson 32)

mile (mi) a customary unit of length; 1 mile = 5,280 feet (Lesson 31)

milliliter (mL) a metric unit of capacity; 1,000 milliliters = 1 liter (Lesson 34)

millimeter (mm) a metric unit of length; 10 millimeters = 1 centimeter (Lesson 32)

minuend in a subtraction problem, the number that is being subtracted from (Lesson 4)

minute (min) a unit of time; 1 minute = 60 seconds (Lesson 36)

month (mo) a period of time; 1 month = 28 to 31 days (Lesson 36)

multiple the product of a number and any other number (Lesson 11)

multiply an operation to find a total when there are equal groups; a shortcut for repeated addition (Lesson 5)

numerator the number above the bar in a fraction; it tells the number of equal parts being considered (Lesson 12)

obtuse angle an angle that has a measure greater than 90° and less than 180° (Lesson 27)

octagon a polygon with 8 sides (Lesson 28)

odd number a whole number that has 1, 3, 5, 7, or 9 in the ones place (Lesson 6)

open sentence a number sentence containing one or more variables; it can be an equation or an inequality (Lesson 22)

parallel describes lines or line segments that stay the same distance apart and never meet (Lesson 26)

parallelogram a quadrilateral with opposite sides parallel (Lesson 28)

pattern an ordered group of numbers or figures that follows a rule (Lesson 24)

pentagon a polygon with 5 sides (Lesson 28)

perimeter the distance around the outside of a closed figure (Lesson 29)

perpendicular describes lines or line segments that intersect at right angles (Lesson 26)

pictograph a graph that shows data by using symbols (Lesson 37)

place value the value of a digit based on its position in a number (Lesson 1)

place-value chart a way of showing numbers that shows the value of each digit (Lesson 1)

point an exact location or position (Lesson 26)

polygon a closed two-dimensional figure with straight sides (Lesson 28)

prediction a statement of what you think will happen (Lesson 40)

product the answer in a multiplication problem (Lesson 5)

quadrilateral a polygon with 4 sides (Lesson 28)

quotient the answer in a division problem (Lesson 9)

ray has an endpoint at one end and goes on forever in the other direction (Lesson 26)

rectangle a parallelogram with 4 right angles (Lesson 28)

rectangular prism a solid figure with 6 faces, 12 edges, and 8 vertices (Lesson 30)

rectangular pyramid a solid figure with 5 faces, 8 edges, and 5 vertices (Lesson 30)

regroup to rename a number for use in addition or subtraction; 15 ones can be regrouped as 1 ten and 5 ones (Lesson 4)

remainder a number less than the divisor that remains after division has ended (Lesson 10)

rhombus a parallelogram with 4 equal sides (Lesson 28)

right angle an angle that has a measure of 90° (Lessons 26, 27)

round to find the nearest value of a number based on a given place value (Lesson 3)

rule an operation that is applied to produce a pattern (Lesson 24)

second (s) a unit of time; 60 seconds = 1 minute (Lesson 36)

sphere a solid figure made up of a one curved surface and shaped like a ball (Lesson 30)

square a rectangle with 4 equal sides (Lesson 28)

standard form a way of writing a number that shows only its digits (Lesson 1)

straight angle an angle with a measure of 180° (Lesson 27)

subtraction an operation to find how many are left after a number is taken away (Lesson 4)

subtrahend the number that is subtracted in a subtraction problem (Lesson 4)

sum the answer in an addition problem (Lesson 4)

survey a way to collect data by asking people questions and recording their answers (Lesson 37)

tally a mark | that represents 1; each ||||| represents 5 (Lessons 37, 39)

three-dimensional figure a figure that has depth; a solid figure (Lesson 30)

trapezoid a quadrilateral with exactly one pair of parallel sides (Lesson 28)

triangle a polygon with 3 sides (Lesson 28)

triangular prism a solid figure with 5 faces, 9 edges, and 6 vertices (Lesson 30)

triangular pyramid a solid figure with 4 triangular faces, 6 edges, and 4 vertices (Lesson 30)

two-dimensional figure a figure that has only length and width (Lesson 28)

unit fraction a fraction with a numerator of 1 (Lesson 14)

variable a letter or symbol used to represent a number (Lesson 22)

vertex (vertices) the point where the rays meet in an angle, where two line segments meet in a polygon, where three or more edges meet in a solid figure, or the point on a cone (Lessons 27, 30)

week (wk) a period of time; 1 week = 7 days (Lesson 36)

whole number any of the numbers 0, 1, 2, 3, and so on (Lesson 1)

word form a way of writing a number using words (Lesson 1)

yard (yd) a customary unit of length; 1 yard = 3 feet (Lesson 31)

year (y) a period of time; 1 year = 12 months (Lesson 36)

New York State Coach, Empire Edition, Mathematics, Grade 4

COMPREHENSIVE REVIEW 1

Name: _____

TIPS FOR TAKING THE TEST

Here are some suggestions to help you do your best:

- Be sure to read carefully all the directions in the test book.
- Read each question carefully and think about the answer before writing your response.
- Be sure to show your work when asked. You may receive partial credit if you have shown your work.

 This picture means that you will use your ruler.

Session 1

1 Pam wrote this pattern in her notebook.

4,392; 5,392; 6,392, _____

What is the next number in Pam's pattern?

A 6,393

B 6,402

C 6,492

D 7,392

2 Which makes this number sentence true?

185 > ____

A 172   C 191

B 186   D 234

3 $\frac{3}{7} + \frac{2}{7} = ?$

A $\frac{1}{7}$ C $\frac{5}{7}$

B $\frac{5}{14}$ D $\frac{6}{7}$

4 Trevor wants to measure the mass of a pen. Which is the **best** unit for Trevor to use?

A gram

B centimeter

C meter

D kilogram

5 The table shows the number of minutes that Kimberly spent on the computer each day from Tuesday to Friday.

KIM'S COMPUTER TIME

Day	Time (in minutes)
Tuesday	35
Wednesday	40
Thursday	45
Friday	50

If the pattern continues, how many minutes will Kimberly spend on the computer on Sunday?

A 40 minutes

B 50 minutes

C 55 minutes

D 60 minutes

6 Which survey question could be answered by observing, rather than asking?

A How many pets do your classmates have?

B What color shirts are your classmates wearing today?

C How many of your classmates have visited the Statue of Liberty?

D What music do the greatest number of your classmates like?

Go On

7 What is the name of this polygon?

A triangle
B quadrilateral
C pentagon
D hexagon

8 Cliff said that his computer cost one thousand, two hundred dollars. Which shows that amount of money?

A $12   C $1,012
B $120   D $1,200

9 Which decimal is equivalent to $\frac{42}{100}$?

A 0.04   C 0.42
B 0.24   D 0.60

10 Nate drew these line segments.

How many more line segments does Nate have to draw to complete an octagon?

A 8   C 5
B 6   D 4

11 LaGarette wants to share candy equally among his 6 friends. If he has 54 pieces of candy, how many pieces will each friend receive?

A 60
B 48
C 9
D 6

12 Liam left for vacation the morning of August 6. He returned the morning of August 19.

AUGUST						
Sun	Mon	Tue	Wed	Thu	Fri	Sat
1	2	3	4	5	6	7
8	9	10	11	12	13	14
15	16	17	18	19	20	21
22	23	24	25	26	27	28
29	30	31				

How many weeks and days was Liam on vacation?

A 1 week 4 days

B 1 week 6 days

C 2 weeks

D 2 weeks 1 day

Go On

13 Andrea wrote this number sentence.

$$0.87 - 0.59 = \Box$$

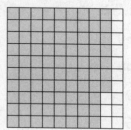

What number belongs in the box to make Andrea's sentence true?

A 0.28   **C** 0.37

B 0.30   **D** 0.38

14 Which sentence is true?

A $8,742 > 8,472$

B $8,472 > 8,724$

C $8,427 > 8,724$

D $8,274 < 8,247$

15 The drawing below shows a stop sign. What is the name of this shape?

A hexagon   **C** pentagon

B octagon   **D** rectangle

16 What decimal represents the shaded portion of this rectangle?

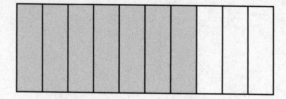

A 0.6

B 0.7

C 0.8

D 0.9

17 Isaac is trying to write a rule that will give him an odd number when he multiplies two numbers. Which rule could Isaac write?

A odd number times even number

B odd number times odd number

C even number times even number

D even number times odd number

18 Daniella is going to measure the length of her paperback book. Which is the **best** unit to use to measure the length of the book?

A inch

B foot

C yard

D mile

Go On

19 The number of books that were borrowed yesterday from the school library is shown in the bar graph below.

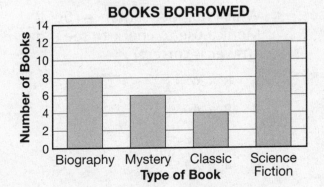

BOOKS BORROWED

Based on the information in the bar graph, which statement is true?

A More science fiction books were borrowed than all other types combined.

B Exactly twice as many science fiction books were borrowed as biography books.

C Exactly twice as many biography books were borrowed as classic books.

D More biography books were borrowed than mystery books and classic books combined.

20 Which point **best** represents $\frac{1}{3}$ on this number line?

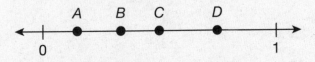

A point *A*   **C** point *C*

B point *B*   **D** point *D*

21 Dominic has 8 more model cars than Ethan. Ethan has 24 model cars. Which open sentence shows how to find how many model cars Dominic has?

A $24 \div 8 = \square$

B $24 - 8 = \square$

C $24 \times 8 = \square$

D $24 + 8 = \square$

22 The Seward Auditorium can seat 838 people. Which is that number rounded to the nearest hundred?

A 800   **C** 840

B 830   **D** 900

23 Which is the **best** estimate for the capacity of a tablespoon?

A 1,000 milliliters

B 500 milliliters

C 300 milliliters

D 15 milliliters

24 Mr. O'Brien got seventy-two cents in change when he bought a sandwich for lunch today. Which is that amount written as a decimal?

A $0.02   **C** $0.72

B $0.70   **D** $7.20

Go On

25 What is the next number in this pattern?

24, 31, 38, 45, _____

A 54   C 52

B 53   D 51

26 The odometer on Mr. Rose's car reads seven thousand, two hundred nine miles. Which is that number of miles written in standard form?

A 7,290   C 7,029

B 7,209   D 729

27 Ms. Connors is 5 feet 6 inches tall. How tall is she in inches?

A 46 inches   C 66 inches

B 56 inches   D 86 inches

28 Cassie is surveying her classmates to find who is their favorite movie star. Which question would be **best** for Cassie to ask to get the information that she wants?

A Who was in the last movie that you went to?

B What is your favorite movie?

C Who is your favorite movie actor or actress?

D What types of movies do you prefer?

29 Monica wants to check that she correctly solved the number sentence below.

32 ÷ 4 = 8

Which number sentence could Monica use to check to see if her answer is correct?

A 8 × 4 = ☐

B 8 ÷ 4 = ☐

C 8 × 32 = ☐

D 4 × 32 = ☐

30 The table shows the scores for four Olympic Gold Medal winners in the heptathlon.

GOLD MEDAL SCORES

Year	Gold Medal Winner	Score
1996	Ghada Shouaa	6,780
2000	Denise Lewis	6,584
2004	Carolina Kluft	6,952
2008	Nataliia Dobrynska	6,733

Which gold medal winner had the **least** score?

A Ghada Shouaa

B Denise Lewis

C Carolina Kluft

D Nataliia Dobrynska

STOP

Session 2

31  Use your ruler to help you solve this problem.

To the nearest $\frac{1}{4}$ inch, how many inches long is the rectangle shown below?

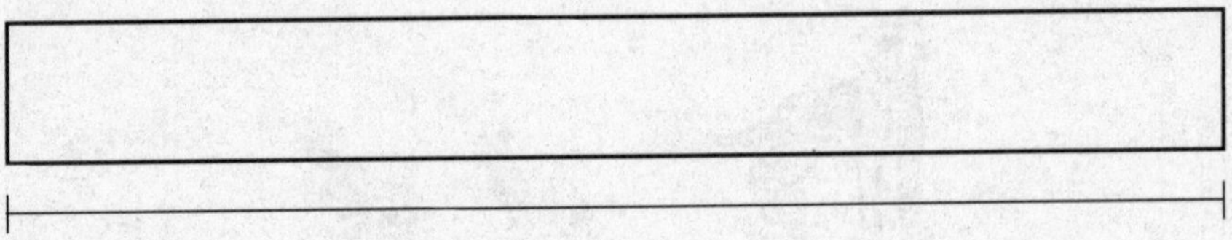

Answer _____ inches

32 The table shows the number of cans that four students collected for recycling yesterday.

CANS COLLECTED

Student	Number of Cans
Sasha	25
Bruce	30
Hal	40
Amy	35

Make a pictograph to show the number of cans that each student collected.

Be sure to

- title the graph
- provide a key for the graph
- graph all the data

KEY
___ = _____

Go On

33 Tierra has 6 pets. Of those pets, $\frac{2}{6}$ are dogs.

Write an equivalent fraction for the number of pets that are dogs.

Answer _____

34 Greg drew this figure and now wants to find the perimeter.

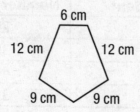

[not drawn to scale]

What is the perimeter of Greg's figure?

Show your work.

Answer _____ centimeters

Go On

35 The table shows the number of people that can ride a roller coaster at the same time.

ROLLER COASTER RIDERS

Number of Cars	Number of People
3	24
4	32
5	40
6	48

Part A

Find the rule of the table.

Show your work.

Answer _____

Part B

How many people can ride in one car of the roller coaster?

Show your work.

Answer _____ people

Go On

36 A total of 22 people signed up to take tennis lessons at the recreational center. 4 people can have lessons every hour. How many hours are needed for everyone to have a lesson?

Show your work.

Answer _____ hours

On the lines below, explain how you interpreted the remainder.

_____

_____

_____

37 Mr. Washington wrote this expression on the board.

$(8 \times 4) \times 5$

What is the product?

Show your work.

Answer _____

Go On

38 Shannon has a box that is shaped like the figure below.

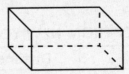

Part A

How many faces does Shannon's box have?

Answer _____ faces

Part B

How many edges does Shannon's box have?

Answer _____ edges

Part C

How many vertices does Shannon's box have?

Answer _____ vertices

Go On

39 The table shows the distance, in miles, of four international cities from New York City.

DISTANCE FROM NEW YORK CITY

City	Distance (in miles)
Berlin, Germany	3,979
Caracas, Venezuela	2,120
Moscow, Russia	4,683
Tokyo, Japan	6,757

Part A

How much closer to New York City is Moscow than Tokyo?

Show your work.

Answer _____ miles

Part B

Darryl is going to fly from New York City to Berlin and back to New York City. How many miles will Darryl fly?

Show your work.

Answer _____ miles

STOP

Session 3

40 Eliza bought a sandwich for $3.69 and paid with a $5.00 bill.

Part A

How much change will she get from the cashier?

Show your work.

Answer $ _____

Part B

On the lines below, give one way the casher can give her that amount of change.

_____

_____

Go On

41 The table shows the number of DVDs that Flo has in her collection.

FLO'S DVD COLLECTION

Type of DVD	Number of DVDs
Comedy	12
Drama	16
Historical	8
Music	10

On the grid below, make a bar graph to show the number of DVDs of each type of DVD Flo has.

Be sure to

• title the graph

• label both axes

• provide a scale for the graph

• graph all the data

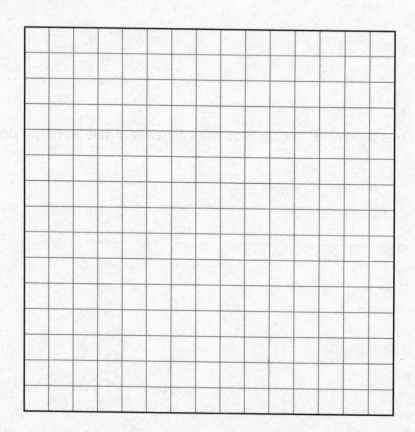

Go On

42 A T-shirt shop sells shirts in packages of 5. Pine Tree Camp needs enough shirts for 68 campers. How many packages of shirts should the camp buy?

Answer _____

On the lines below, explain your answer.

_____

_____

_____

_____

43 The diagram represents the area of Heidi's kitchen.

	KEY
☐ = 1 square foot

What is the area of Heidi's kitchen?

Answer _____ square feet

Go On

44 Rahim pays $37 each month for his cell phone plan. How much money will Rahim spend in 6 months for his cell phone plan?

Show your work.

Answer $ _____

45 Dara's soccer practice began at 7:30 A.M. Practice ended at 10:00 A.M. How many hours and minutes did soccer practice last?

Answer _____

46 Each package of loose-leaf paper that Robin bought contains 100 sheets. She bought 5 packages. How many sheets of loose-leaf paper did Robin buy?

Show your work.

Answer _____ sheets

Go On

47 Frank receives an allowance. He puts a certain amount of his allowance in a savings account each week. The table below shows how much money Frank has saved each week.

FRANK'S SAVINGS

Week	Money Saved
1	$5
2	$10
3	$15
4	$20

Part A

If the pattern in the table continues, how much money will Frank have saved in week 6?

Answer $ _____

Part B

On the lines below, explain the rule to find the number of dollars that Frank saves.

_____

_____

_____

_____

_____

Go On

48 Tina surveyed her classmates to find which subject other than gym is their favorite. Her results are shown below.

math, reading, math, social studies, math, science,

math, reading, social studies, math, reading, reading,

math, science, science, math, science, social studies,

reading, math, reading, reading, science, math

Part A

Complete the tally table to show the data.

FAVORITE SUBJECTS

Subject	Tally	Number

Part B

On the grid to the right, make a bar graph to show the students' favorite subjects.

Be sure to

• title the graph

• label both axes

• provide a scale for the graph

• graph all the data

STOP

New York State Coach, Empire Edition, Mathematics, Grade 4

COMPREHENSIVE REVIEW 2

Name: _____

TIPS FOR TAKING THE TEST

Here are some suggestions to help you do your best:

- Be sure to read carefully all the directions in the test book.

- Read each question carefully and think about the answer before writing your response.

- Be sure to show your work when asked. You may receive partial credit if you have shown your work.

 This picture means that you will use your ruler.

Session 1

1 Which describes the shaded part of this rectangle?

- **A** $\frac{1}{6}$ and $\frac{1}{3}$
- **B** $\frac{2}{6}$ and $\frac{1}{3}$
- **C** $\frac{2}{6}$ and $\frac{1}{2}$
- **D** $\frac{4}{6}$ and $\frac{2}{3}$

2 With only one thousand, forty-five square miles of land area, Rhode Island is the smallest state in the United States. Which is Rhode Island's land area written in expanded form?

- **A** $100 + 40 + 5$
- **B** $1,000 + 40 + 5$
- **C** $1,000 + 400 + 5$
- **D** $1,000 + 400 + 50$

3 Which is the **best** estimate for the mass of a baseball bat?

- **A** 10 grams
- **B** 100 grams
- **C** 1,000 grams
- **D** 5,000 grams

4 Deanna is going to donate books to a library. She has 6 boxes and put 18 books into each box. Which open sentence shows how to find how many books Deanna will donate?

- **A** $6 \times 18 = \underline{\qquad}$
- **B** $6 + 18 = \underline{\qquad}$
- **C** $18 - 6 = \underline{\qquad}$
- **D** $18 \div 6 = \underline{\qquad}$

5 Each raffle ticket costs 5 dollars. Each book of raffle tickets contains 20 tickets. Brianna sold 4 books of raffle tickets. Which expression could be used to find the total amount of money that Brianna collected?

- **A** $5 \times 20 + 4$
- **B** $5 \times 20 \times 4$
- **C** $5 + 20 + 4$
- **D** $5 + 20 \times 4$

6 What is the product of 35×18?

- **A** 280
- **B** 590
- **C** 630
- **D** 6,300

Go On

7 Sandra recorded the temperature outside her house every day at 5 P.M. The graph below shows her data for the beginning of May.

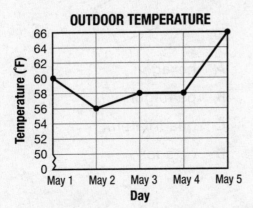

OUTDOOR TEMPERATURE

Between which two days did the temperature increase the most?

A May 1 and 2

B May 2 and 3

C May 3 and 4

D May 4 and 5

8 The two statements below describe the number of Mike's, Jose's, and Karen's in Ms. Jackson's class.

- the number of Jose's > the number of Karen's

- the number of Mike's = the number of Jose's

Which could be the number of Mike's, Jose's, and Karen's in Ms. Jackson's class?

A 3 Jose's, 3 Karen's, 2 Mike's

B 2 Jose's, 2 Mike's, 3 Karen's

C 2 Karen's, 2 Mike's, 3 Jose's

D 2 Mike's, 2 Jose's, 1 Karen

9 The number of minutes that Alan spent on the computer this week is shown in the bar graph below.

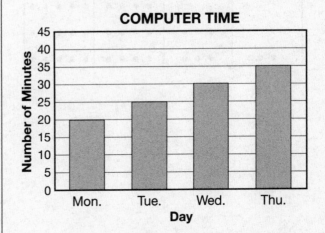

COMPUTER TIME

Based on the information in the bar graph, how much time do you predict Alan will spend on the computer on Friday?

A 40 minutes

B 35 minutes

C 30 minutes

D 25 minutes

10 Connor wrote the number sentence below.

$$27 \div 3 = 9$$

Which number sentence could Connor use to check if his answer is correct?

A $27 \times 3 = \square$

B $27 \times 9 = \square$

C $9 \div 3 = \square$

D $9 \times 3 = \square$

Go On

11 How many dots are in the next figure in this pattern?

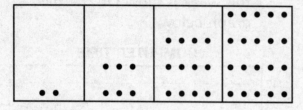

- A 26
- B 28
- C 30
- D 32

12 Adam wants to measure 2 milliliters of food coloring for a recipe. Which tool should he use to measure the food coloring?

- A ruler
- B scale
- C thermometer
- D dropper

13 Which number, when multiplied by 3, results in an odd number?

- A 2
- B 4
- C 6
- D 9

14 What is the name of the shape below?

- A hexagon
- B pentagon
- C quadrilateral
- D octagon

15 Tatiana wants to measure the length of her bedroom. Which is the **best** tool for Tatiana to use?

- A tape measure
- B protracter
- C ruler
- D scale

16 Benjamin is going to his grandparents' house and then to his uncle's house. Benjamin lives 332 miles from his grandparents. His grandparents live 189 miles from his uncle. Which expression is **best** to estimate the number of miles Benjamin will travel to get to his uncle's house?

- A 300 + 100
- B 330 + 100
- C 330 + 190
- D 400 + 200

Go On

17 One mile is equal to one thousand, seven hundred sixty yards. What is this number in expanded form?

A 10,000 + 700 + 60

B 1,000 + 700 + 60

C 1,000 + 700 + 10 + 6

D 100 + 70 + 6

18 Ernie is surveying his classmates to find their favorite sport to watch on television. Which question would be **best** for Ernie to ask to get the information that he wants?

A Which sports do you play?

B Which sport are you the best at playing?

C Which sport did you last watch on television?

D Which sport do you most like watching on television?

19 The longest roller coaster in the United States is 7,400 feet long. How is that number written in word form?

A seventy-four thousand

B seven thousand, four hundred

C seven thousand, four

D seven hundred forty

20 The table shows the altitudes of the highest points in four U.S. states.

HIGHEST POINTS

State	Altitude (in feet)
Maine	5,268
Nebraska	5,424
New York	5,344
Virginia	5,729

Which state's highest point is **greatest**?

A Maine

B Nebraska

C New York

D Virginia

21 Which number belongs on the line below to make the number sentence correct?

$$6 + \underline{} > 5 + 7$$

A 7

B 6

C 5

D 4

Go On

22 Bianca is drawing a hexagon. She has drawn the figure below.

How many more line segments does Bianca need to draw to complete the figure?

A   4

B   3

C   2

D   1

23 The number of CDs that Doug has uploaded to his computer is shown in the table.

DOUG'S CD UPLOADS

Week	Number of Uploads
2	18
4	36
6	54
8	72
10	?

If the pattern continues, how many CDs will Doug have uploaded to his computer by the end of Week 10?

A   81

B   90

C   100

D   126

24 It takes 10 yards to earn a first down in football. How many feet does it take to earn a first down?

A   360 feet

B   120 feet

C   30 feet

D   20 feet

25 Which number is equal to 6 thousands, 2 tens, and 5 ones?

A   625   C   6,205

B   6,025   D   6,250

26 A truck driver is planning a trip. He will leave Albany the morning of November 6 and return the morning of November 24.

NOVEMBER						
Sun	Mon	Tue	Wed	Thu	Fri	Sat
	1	2	3	4	5	6
7	8	9	10	11	12	13
14	15	16	17	18	19	20
21	22	23	24	25	26	27
28	29	30				

How many days and weeks is his trip?

A   3 weeks 4 days

B   3 weeks 1 day

C   2 weeks 4 days

D   2 weeks 1 day

Go On

27 Which sentence is true?

A $2,803 > 2,083$

B $3,725 \ 3,725$

C $4,106 = 4,160$

D $6,382 < 6,238$

28 Mr. Dylan wants to make 8 fluid ounces of punch for each of his 18 guests at a party. He said he needs to make about 160 fluid ounces of punch. Which statement **best** explains whether Mr. Dylan's estimate is reasonable?

A It is reasonable because $8 \times 20 = 160$.

B It is reasonable because $10 \times 18 = 180$.

C It is not reasonable because $8 \times 18 = 144$.

D It is not reasonable because $8 \times 10 = 80$.

29 Stan is trying to explain multiplication to his younger sister. Which phrase **best** explains the meaning of multiplication?

A taking away some from a group

B joining groups of different sizes

C sharing in equal numbers from a group

D joining groups that are all the same size

30 How many line segments does the shape shown below have?

A 5

B 6

C 7

D 8

STOP

Session 2

31 *Part A*

Each expression in the first column below is equivalent to one of the expressions in the second column. Draw a line between the pairs of expressions that are equivalent. The first line has been drawn for you.

$5 \times (4 \times 6)$ $3 + 11$

$(5 + 4) + 6$ $3 \times (2 \times 8)$

$(3 \times 2) \times 8$ $(5 \times 4) \times 6$

$3 + (3 + 8)$ $9 + 6$

Part B

Brett wrote the number sentence below:

$(2 \times 3) \times 5 = 30$

Hannah rewrote Brett's number sentence and grouped the numbers differently. Complete Hannah's number sentence below to show the new grouping. Use **all** the numbers and symbols from Joseph's number sentence.

Hannah's number sentence _____ $= 30$

Go On

32 Nia makes the Input-Output table below.

NIA'S INPUT-OUTPUT TABLE

Input	21	28	35	42
Output	3	4	5	6

Part A

What rule could be used to find each Output number?

Answer _____

Part B

If the next two Input numbers are 56 and 63, what would the Output numbers be?

Answer _____

33 The diagram represents the area of Ralph's basement.

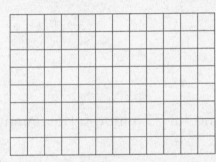

KEY
☐ = 1 square yard

What is the area of Ralph's basement?

Answer _____ square yards

Go On

34 Marcy bought a magazine for $2.59 and a carton of juice for $1.25. She paid with a $10.00 bill. How much change should Marcy receive?

Show your work.

Answer $ _____

35 For a birthday party, Ms. Lee made a chocolate cake and an apple pie. After the party, she had $\frac{1}{4}$ of the cake and $\frac{3}{4}$ of the pie left. On the number line below, show the fraction of the cake and the fraction of the pie she had left.

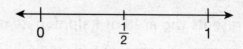

36 Each day, Diana practices the piano for 75 minutes. How many minutes does she practice in 7 days?

Show your work.

Answer _____ minutes

Go On

37 A total of 84 people volunteered to help clean up a state park. They are going to be divided into 3 equal groups. How many people will be in each group?

Show your work.

Answer _____ people

38 Look at the diagram below to answer the questions.

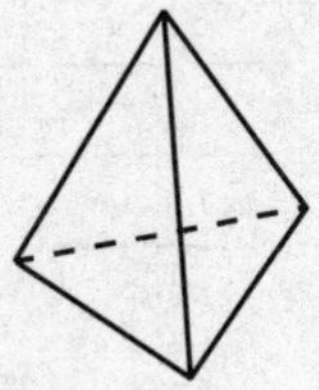

Part A

How many faces does the figure have?

Answer _____ faces

Part B

How many edges does the figure have?

Answer _____ edges

Part C

How many vertices does the figure have?

Answer _____ vertices

Go On

39 Jake surveyed his classmates to find which type of music is their favorite. Their answers are shown below.

> rock, hip hop, classical, rock, rock, hip hop, country, hip hop,
>
> country, hip hop, rock, country, hip hop, hip hop, rock, rock,
>
> country, rock, rock, classical, rock, country, country, hip hop

Part A

Complete the tally table to show the data.

FAVORITE TYPES OF MUSIC

Music Type	Tally	Number

Part B

On the grid, make a bar graph to show the data in the tally table.

Be sure to

- title the graph
- label both axes
- provide a scale for the graph
- graph all the data

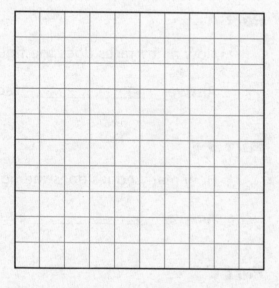

STOP

Session 3

40 Dexter began writing a paper at 4:30 P.M. He finished writing at 6:00 P.M. How many hours and minutes did it take Dexter to write his paper?

Show your work.

Answer _____

41 Of the shapes shown below, $\frac{2}{8}$ are stars.

Write an equivalent fraction for the fraction of shapes that are stars.

Answer _____

42 It takes Rosemary and Michelle 10 minutes to walk 1 lap around Madison Park. How many minutes will it take the girls to walk 8 laps around the park?

Show your work.

Answer _____ minutes

Go On

43 The table shows the number of points that the Lions basketball team scored in its first four games.

POINTS SCORED

Opponent	Points Scored
East	28
West	20
North	36
South	40

Make a pictograph to show the number of points that the Lions scored in each of their games.

Be sure to

- title the graph

- provide a key for the graph

- graph all the data

KEY
___ = _____

Go On

44 What is the perimeter of this figure?

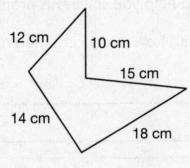

[not drawn to scale]

Show your work.

Answer _____ centimeters

45 Kris baked 44 cookies. She wants to put them in bags, with 6 cookies in each bag. How many bags can she fill?

Show your work.

Answer _____ bags

Go On

46 Use your ruler to help you solve this problem.

To the nearest centimeter, how many centimeters long is the rectangle shown below?

Answer _____ centimeters

Go On

47 The table shows the lengths of four bridges in New York.

NEW YORK BRIDGES

Bridge	Length (in feet)
Verrazano-Narrows	4,260
Throgs Neck	1,801
Bayonne	1,675
Tappan Zee	1,212

Part A

How much longer is the Verrazano-Narrows Bridge than the Bayonne Bridge?

Show your work.

Answer _____ feet

Part B

How much longer is the Verrazano-Narrows Bridge than the Throgs Neck Bridge and the Tappan Zee Bridge combined?

Show your work.

Answer _____ feet

Go On

48 Laura is saving money to buy a new netbook. Her aunt gives her a certain amount of money for each dollar Laura saves. The table below shows how much money Laura has saved and how much money her aunt has given her.

LAURA'S SAVINGS

Laura	Aunt
$5	$20
$6	$24
$7	$28
$8	$32

Part A

If the pattern in the table continues, how much money will Laura's aunt have given her when Laura saves $10?

Answer $ _____

Part B

On the lines below, explain the rule to find the number of dollars Laura's aunt gives her.

_____

_____

_____

_____

_____

STOP

Punch-Out Tools

Notes